STUDENT'S SOLUTIONS MANUAL

EDGAR REYES

Southeastern Louisiana University

TRIGONOMETRY
FOURTH EDITION

Mark Dugopolski

Southeastern Louisiana University

Boston Columbus Indianapolis New York San Francisco Upper Saddle River
Amsterdam Cape Town Dubai London Madrid Milan Munich Paris Montreal Toronto
Delhi Mexico City São Paulo Sydney Hong Kong Seoul Singapore Taipei Tokyo

Copyright © 2015, 2011, 2007 Pearson Education, Inc.
Publishing as Pearson, 75 Arlington Street, Boston, MA 02116.

ISBN-13: 978-0-321-91547-4
ISBN-10: 0-321-91547-X

1 2 3 4 5 6 EBM 17 16 15 14

www.pearsonhighered.com

PEARSON

Table of Contents

Chapter P..1

Chapter 1..18

Chapter 2...43

Chapter 3...71

Chapter 4..99

Chapter 5...125

Chapter 6...154

For Thought

1. False, the point $(2, -3)$ is in Quadrant IV.

2. False, the point $(4, 0)$ does not belong to any quadrant.

3. False, since the distance is $\sqrt{(a-c)^2 + (b-d)^2}$.

4. False, since $Ax + By = C$ is a linear equation.

5. True

6. False, since $\sqrt{7^2 + 9^2} = \sqrt{130} \approx 11.4$

7. True

8. True

9. True

10. False, since the radius is 3.

P.1 Exercises

1. ordered

3. x-axis

5. Pythagorean theorem

7. linear equation

9. $(4, 1)$, Quadrant I

11. $(1, 0)$, x-axis

13. $(5, -1)$, Quadrant IV

15. $(-4, -2)$, Quadrant III

17. $(-2, 4)$, Quadrant II

19. $c = \sqrt{(\sqrt{3})^2 + 1^2} = \sqrt{4} = 2$

21. Since $b^2 + 2^2 = 3^2$, we get $b^2 + 4 = 9$ or $b^2 = 5$. Then $b = \sqrt{5}$.

23. Since $a^2 + 3^2 = 5^2$, we get $a^2 + 9 = 25$ or $a^2 = 16$. Then $a = 4$.

25. $\sqrt{4 \cdot 7} = 2\sqrt{7}$

27. $\dfrac{\sqrt{5}}{\sqrt{9}} = \dfrac{\sqrt{5}}{3}$

29. $\dfrac{\sqrt{2}}{\sqrt{3}} \cdot \dfrac{\sqrt{3}}{\sqrt{3}} = \dfrac{\sqrt{6}}{3}$

31. $\dfrac{2\sqrt{3}}{\sqrt{5}} \cdot \dfrac{\sqrt{5}}{\sqrt{5}} = \dfrac{2\sqrt{15}}{5}$

33. $\dfrac{1}{\sqrt{3}} \cdot \dfrac{\sqrt{3}}{\sqrt{3}} = \dfrac{\sqrt{3}}{3}$

35. $\dfrac{\sqrt{2}}{\sqrt{3}} \cdot \dfrac{\sqrt{3}}{\sqrt{3}} = \dfrac{\sqrt{6}}{3}$

37. Distance is $\sqrt{(4-1)^2 + (7-3)^2} = \sqrt{9+16} = \sqrt{25} = 5$, midpoint is $(2.5, 5)$

39. Distance is $\sqrt{(-1-1)^2 + (-2-0)^2} = \sqrt{4+4} = 2\sqrt{2}$, midpoint is $(0, -1)$

41. Distance is $\sqrt{\left(\dfrac{\sqrt{2}}{2} - 0\right)^2 + \left(\dfrac{\sqrt{2}}{2} - 0\right)^2} = \sqrt{\dfrac{2}{4} + \dfrac{2}{4}} = \sqrt{1} = 1$, midpoint is $\left(\dfrac{\sqrt{2}/2 + 0}{2}, \dfrac{\sqrt{2}/2 + 0}{2}\right) = \left(\dfrac{\sqrt{2}}{4}, \dfrac{\sqrt{2}}{4}\right)$

43. Distance is $\sqrt{\left(\sqrt{18} - \sqrt{8}\right)^2 + \left(\sqrt{12} - \sqrt{27}\right)^2} = \sqrt{(3\sqrt{2} - 2\sqrt{2})^2 + (2\sqrt{3} - 3\sqrt{3})^2} = \sqrt{(\sqrt{2})^2 + (-\sqrt{3})^2} = \sqrt{5}$, midpoint is $\left(\dfrac{\sqrt{18} + \sqrt{8}}{2}, \dfrac{\sqrt{12} + \sqrt{27}}{2}\right) = \left(\dfrac{3\sqrt{2} + 2\sqrt{2}}{2}, \dfrac{2\sqrt{3} + 3\sqrt{3}}{2}\right) = \left(\dfrac{5\sqrt{2}}{2}, \dfrac{5\sqrt{3}}{2}\right)$

45. Distance is $\sqrt{(1.2 + 3.8)^2 + (4.4 + 2.2)^2} = \sqrt{25 + 49} = \sqrt{74}$, midpoint is $(-1.3, 1.3)$

47. Distance is $\dfrac{\sqrt{\pi^2 + 4}}{2}$, midpoint is $\left(\dfrac{3\pi}{4}, \dfrac{1}{2}\right)$

49. Distance is $\sqrt{(2\pi - \pi)^2 + (0 - 0)^2} = \sqrt{\pi^2} = \pi$, midpoint is $\left(\dfrac{2\pi + \pi}{2}, \dfrac{0 + 0}{2}\right) = \left(\dfrac{3\pi}{2}, 0\right)$

51. Center $(0, 0)$, radius 4

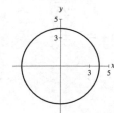

53. Center $(-6, 0)$, radius 6

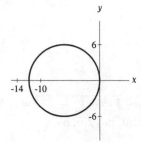

55. Center $(2, -2)$, radius $2\sqrt{2}$

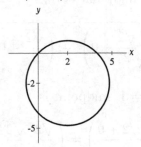

57. $x^2 + y^2 = 7$

59. $(x + 2)^2 + (y - 5)^2 = 1/4$

61. The distance between $(3, 5)$ and the origin is $\sqrt{34}$ which is the radius. The standard equation is $(x - 3)^2 + (y - 5)^2 = 34$.

63. Note, the distance between $(\sqrt{2}/2, \sqrt{2}/2)$ and the origin is 1. Thus, the radius is 1. The standard equation is $x^2 + y^2 = 1$.

65. The radius is $\sqrt{(-1 - 0)^2 + (2 - 0)^2} = \sqrt{5}$. The standard equation is $(x+1)^2 + (y-2)^2 = 5$.

67. Note, the center is $(1, 3)$ and the radius is 2. The standard equation is $(x-1)^2 + (y-3)^2 = 4$.

69. We solve for a.

$$a^2 + \left(\frac{3}{5}\right)^2 = 1$$
$$a^2 = 1 - \frac{9}{25}$$
$$a^2 = \frac{16}{25}$$
$$a = \pm\frac{4}{5}$$

71. We solve for a.

$$\left(-\frac{2}{5}\right)^2 + a^2 = 1$$
$$a^2 = 1 - \frac{4}{25}$$
$$a^2 = \frac{21}{25}$$
$$a = \pm\frac{\sqrt{21}}{5}$$

73. $y = 3x - 4$ goes through $(0, -4)$, $\left(\frac{4}{3}, 0\right)$.

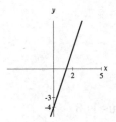

75. $3x - y = 6$ goes through $(0, -6)$, $(2, 0)$.

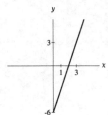

77. $x + y = 80$ goes through $(0, 80)$, $(80, 0)$.

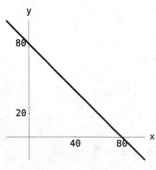

79. $x = 3y - 90$ goes through $(0, 30)$, $(-90, 0)$.

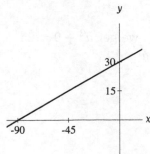

81. $\frac{1}{2}x - \frac{1}{3}y = 600$ goes through $(0, -1800)$, $(1200, 0)$.

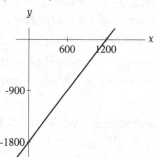

83. Intercepts are $(0, 0.0025), (0.005, 0)$.

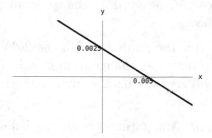

85. $x = 5$

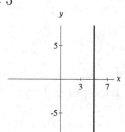

87. $y = 4$

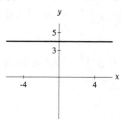

89. $x = -4$

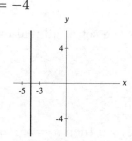

91. Solving for y, we have $y = 1$.

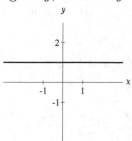

93. $y = x - 20$ goes through $(0, -20)$, $(20, 0)$.

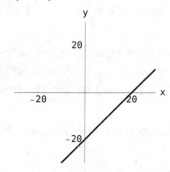

95. $y = 3000 - 500x$ goes through $(0, 3000)$, $(6, 0)$.

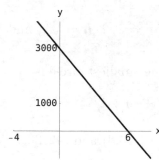

97. The hypotenuse is $\sqrt{6^2 + 8^2} = \sqrt{100} = 10$.

99. a) Let r be the radius of the smaller circle. Consider the right triangle with vertices at the origin, another vertex at the center of a smaller circle, and a third vertex at the center of the circle of radius 1. By the Pythagorean Theorem, we obtain

$$
\begin{aligned}
1 + (2 - r)^2 &= (1 + r)^2 \\
5 - 4r &= 1 + 2r \\
4 &= 6r \\
r &= \frac{2}{3}.
\end{aligned}
$$

The diameter of the smaller circle is $2r = \frac{4}{3}$.

b) The smallest circles are centered at $(\pm r, 0)$ or $(\pm 4/3, 0)$. The equations of the circles are

$$\left(x - \frac{4}{3}\right)^2 + y^2 = \frac{4}{9}$$

and

$$\left(x + \frac{4}{3}\right)^2 + y^2 = \frac{4}{9}$$

101. Let $C(h, k)$ and r be the center and radius of the smallest circle, respectively. Then $k = -r$. We consider two right triangles each of which has a vertex at C.

The right triangles have sides that are perpendicular to the coordinate axes. Also, one side of each right triangle passes through the center of a larger circle.

Applying the Pythagorean Theorem, we list a system of equations

$$(r + 1)^2 = h^2 + (1 - r)^2$$
$$(2 - r)^2 = h^2 + r^2.$$

The solutions are $r = 1/2$, $h = \sqrt{2}$, and $k = -r = -1/2$.

The equation of the smallest circle is

$$(x - \sqrt{2})^2 + (y + 1/2)^2 = 1/4.$$

103. The midpoint of $(0, 20.8)$ and $(42, 26.1)$ is

$$\left(\frac{0 + 42}{2}, \frac{20.8 + 26.1}{2}\right) \approx (21, 23.5).$$

In 1991 $(= 1970 + 21)$, the median age at first marriage is 23.5 years

105. The distance between $(10, 0)$ and $(0, 0)$ is 10. The distance between $(1, 3)$ and the origin is $\sqrt{10}$. If two points have integer coordinates, then the distance between them is of the form $\sqrt{s^2 + t^2}$ where $s^2, t^2 \in \{0, 1, 2^2, 3^2, 4^2, ...\} = \{0, 1, 4, 9, 16, ...\}$.

Note, there are no numbers s^2 and t^2 in $\{0, 1, 4, 9, 16, ...\}$ satisfying $s^2 + t^2 = 19$. Thus, one cannot find two points with integer coordinates whose distance between them is $\sqrt{19}$.

109. On day 1, break off a 1-dollar piece and pay the gardener.

On day 2, break off a 2-dollar and pay the gardener. The gardener will give you back your change which is a 1-dollar piece.

On day 3, you pay the gardener with the 1-dollar piece you received as change from the previous day.

On day 4, pay the gardener with the 4-dollar bar. The gardener will give you back your change which will consist of a 1-dollar piece and a 2-dollar piece.

On day 5, you pay the gardener with the 1-dollar piece you received as change from the previous day.

On day 6, pay the gardener with the 2-dollar piece you received as change from day 4. The gardener will give you back your change which is a 1-dollar piece.

On day 7, pay the gardener with the 1-dollar piece you received as change from day 6.

For Thought

1. True, since the number of gallons purchased is 20 divided by the price per gallon.

2. False, since a student's exam grade is a function of the student's preparation. If two classmates had the same IQ and only one prepared then the one who prepared will most likely achieve a higher grade.

3. False, since $\{(1, 2), (1, 3)\}$ is not a function.

4. True

5. True

6. True

7. False, the domain is the set of all real numbers.

8. True

9. True, since $f(0) = \dfrac{0 - 2}{0 + 2} = -1$.

10. True, since if $a - 5 = 0$ then $a = 5$.

P.2 Exercises

1. function

3. domain, range

5. function

7. Note, $b = 2\pi a$ is equivalent to $a = \dfrac{b}{2\pi}$. Then a is a function of b, and b is a function of a.

9. a is a function of b since a given denomination has a unique length. Since a dollar bill and a five-dollar bill have the same length, then b is not a function of a.

11. Since an item has only one price, b is a function of a. Since two items may have the same price, a is not a function of b.

13. a is not a function of b since it is possible that two different students can obtain the same final exam score but the times spent on studying are different.

b is not a function of a since it is possible that two different students can spend the same time studying but obtain different final exam scores.

15. Since 1 in ≈ 2.54 cm, a is a function of b and b is a function of a.

17. Since $b = a^3$ and $a = \sqrt[3]{b}$, we get that b is a function of a, and a is a function of b.

19. Since $b = |a|$, we get b is a function of a. Since $a = \pm b$, we find a is not a function of b.

21.   **23.** $s = \dfrac{\sqrt{2}d}{2}$ **25.** $P = 4s$

27. $A = P^2/16$

29. $y = 2x - 1$ has domain $(-\infty, \infty)$ and range $(-\infty, \infty)$, some points are $(0, -1)$ and $(1, 1)$

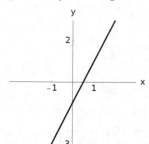

31. $y = 5$ has domain $(-\infty, \infty)$ and range $\{5\}$, some points are $(0, 5)$ and $(1, 5)$

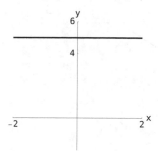

33. $y = x^2 - 20$ has domain $(-\infty, \infty)$ and range $[-20, \infty)$, some points are $(0, -20)$ and $(6, 16)$

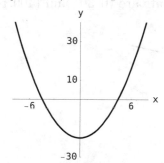

35. $y = 40 - x^2$ has domain $(-\infty, \infty)$ and range $(-\infty, 40]$, some points are $(0, 40)$ and $(6, 4)$

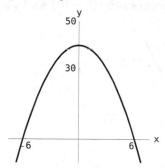

37. $y = x^3$ has domain $(-\infty, \infty)$ and range $(-\infty, \infty)$, some points are $(0, 0)$ and $(2, 8)$

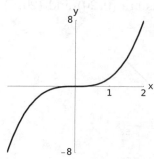

39. $y = \sqrt{x-10}$ has domain $[10, \infty)$ and range $[0, \infty)$, some points are $(10, 0)$ and $(14, 2)$

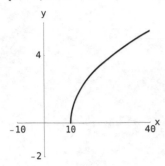

41. $y = \sqrt{x} + 30$ has domain $[0, \infty)$ and range $[30, \infty)$, some points are $(0, 30)$ and $(400, 50)$

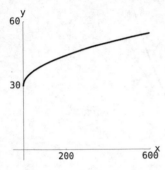

43. $y = |x| - 40$ has domain $(-\infty, \infty)$ and range $[-40, \infty)$, some points are $(0, -40)$ and $(40, 0)$

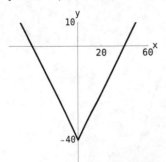

45. $y = |x - 20|$ has domain $(-\infty, \infty)$ and range $[0, \infty)$, some points are $(0, 20)$ and $(20, 0)$

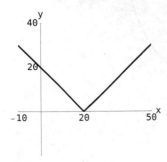

47. $3 \cdot 4 - 2 = 10$

49. $-4 - 2 = -6$

51. $|8| = 8$

53. $0.9408 - 0.56 = 0.3808$

55. $4 + (-6) = -2$

57. $80 - 2 = 78$

59. $3a^2 - a$

61. $f(-x) = 3(-x)^2 - (-x) = 3x^2 + x$

63. Factoring, we get $x(3x - 1) + 0$. So $x = 0, 1/3$.

65. Since $|a + 3| = 4$ is equivalent to $a + 3 = 4$ or $a + 3 = -4$, we have $a = 1, -7$.

67. $C = 353n$

69. $C = 35n + 50$

71. We find
$$C = \frac{4B}{\sqrt[3]{D}} = \frac{4(12 + 11/12)}{\sqrt[3]{22,800}} \approx 1.822$$

and a sketch of the graph of $C = \dfrac{4B}{\sqrt[3]{22,800}}$ is given below.

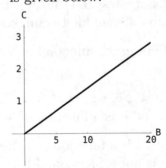

73. Let $N = 2$, $B = 3.498$, and $S = 4.250$. Then
$$
\begin{aligned}
D &= \frac{\pi}{4}B^2 \cdot S \cdot N \\
&= 81.686 \text{ in.}^3
\end{aligned}
$$
Then $D \approx 81.7$ in.^{3}.

75. Solving for B,
$$
\begin{aligned}
D &= \frac{\pi}{4}B^2 \cdot S \cdot N \\
\frac{4D}{\pi S \cdot N} &= B^2 \\
B &= 2\sqrt{\frac{D}{\pi S \cdot N}}.
\end{aligned}
$$

77. Pythagorean, legs, hypotenuse

79 . $\sqrt{(2+3)^2 + (-4+6)^2} = \sqrt{29}$

81. If we replace $x = 0$ in $4x - 6y = 40$, then $-6y = 40$ or $y = -20/3$. The y-intercept is $(0, -\frac{20}{3})$.

If we replace $y = 0$ in $4x - 6y = 40$, then $4x = 40$ or $x = 10$. The x-intercept is $(10, 0)$.

83. Rewriting the equation, we find

$$\frac{1}{3^3} \cdot 3^{100} \cdot \frac{1}{3^4} \cdot 3^{2x} = \frac{1}{3} \cdot 3^x$$
$$\frac{1}{3^3} \cdot 3^{100} \cdot \frac{1}{3^4} \cdot 3^{2x} = 3^{x-1}$$
$$3^{2x+93} = 3^{x-1}$$
$$2x + 93 = x - 1$$
$$x = -94.$$

For Thought

1. False, it is a reflection in the y-axis.

2. False, the graph of $y = x^2 - 4$ is shifted down 4 units from the graph of $y = x^2$.

3. False, rather it is a left translation.

4. True

5. True

6. False, the down shift should come after the reflection.

7. True

8. False, since the domains are different.

9. True

10. True, since $f(-x) = -f(x)$ where $f(x) = x^3$.

P.3 Exercises

1. rigid

3. reflection

5. right, left

7. odd

9. transformation

11. $f(x) = \sqrt{x}$, $g(x) = -\sqrt{x}$

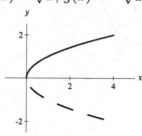

13. $y = x$, $y = -x$

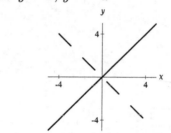

15. $f(x) = |x|$, $g(x) = |x| - 4$

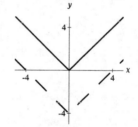

17. $f(x) = x$, $g(x) = x + 3$

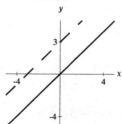

19. $y = x^2$, $y = (x-3)^2$

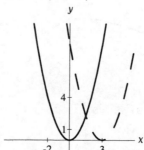

21. $f(x) = x^3$, $g(x) = (x+1)^3$

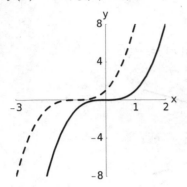

23. $y = \sqrt{x}$, $y = 3\sqrt{x}$

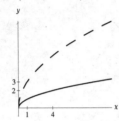

25. $y = x^2$, $y = \dfrac{1}{4}x^2$

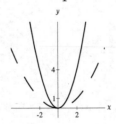

27. g   **29.**   b

31. c   **33.**   f

35. $y = (x-10)^2 + 4$

37. $y = -3|x-7| + 9$

39. $y = -(3\sqrt{x} + 5)$ or $y = -3\sqrt{x} - 5$

41. $y = \sqrt{x-1} + 2$; right by 1, up by 2

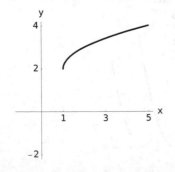

43. $y = |x-1| + 3$; right by 1, up by 3

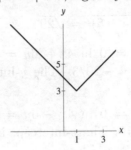

45. $y = 3x - 40$

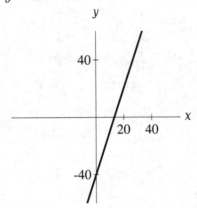

47. $y = \dfrac{1}{2}x - 20$

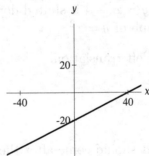

49. $y = -\dfrac{1}{2}|x| + 40$, shrink by $1/2$,

reflect about x-axis, up by 40

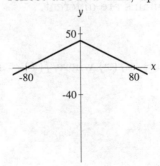

51. $y = -\dfrac{1}{2}|x+4|$, left by 4,

reflect about x-axis, shrink by 1/2

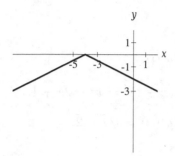

53. $y = -(x-3)^2 + 1$; right by 3,
reflect about x-axis, up by 1

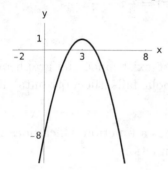

55. $y = -2(x+3)^2 - 4$; left by 3,
stretch by 2, reflect about x-axis, down by 4

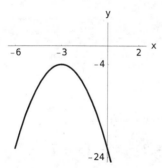

57. $y = -2\sqrt{x+3} + 2$, left by 3, stretch by 2,
reflect about x-axis, up by 2

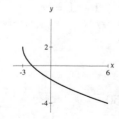

59. Symmetric about y-axis, even function
since $f(-x) = f(x)$

61. No symmetry, neither even nor odd
since $f(-x) \neq f(x)$ and $f(-x) \neq -f(x)$

63. Neither symmetry, neither even nor odd since
$f(-x) \neq f(x)$ and $f(-x) \neq -f(x)$

65. No symmetry, not an even or odd function
since $f(-x) = -f(x)$ and $f(-x) \neq -f(x)$

67. Symmetric about the origin, odd function
since $f(-x) = -f(x)$

69. No symmetry, not an even or odd function
since $f(-x) \neq f(x)$ and $f(-x) \neq -f(x)$

71. No symmetry, not an even or odd function
since $f(-x) \neq f(x)$ and $f(-x) \neq -f(x)$

73. Neither symmetry, not an even or odd function
since $f(-x) \neq f(x)$ and $f(-x) \neq -f(x)$

75. Symmetric about the y-axis, even function
since $f(-x) = f(x)$

77. Symmetric about the y-axis, even function
since $f(-x) = f(x)$

79. e   **81.**   g

83. b   **85.**   c

87. $N(x) = x + 2000$

89. If inflation rate is less than 50%, then

$1 - \sqrt{x} < \dfrac{1}{2}$. This simplifies to $\dfrac{1}{2} < \sqrt{x}$. After

squaring we have $\dfrac{1}{4} < x$ and so $x > 25\%$.

91.

   (a) Both functions are symmetric about the
y-axis, and the graphs are identical.

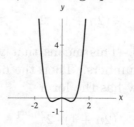

(b) One graph is a reflection of the other about the y-axis, and both are symmetric about the y-axis.

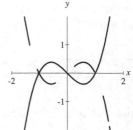

(c) The second graph is obtained by translating the first one to the left by 1 unit.

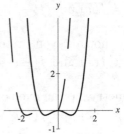

(d) The second graph is obtained by translating the first one to the right by 2 units and 3 units up.

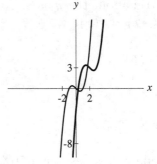

93. $x^2 + y^2 = 1$

95. $x = 4$

97. $f(6) = 2(36) - 3(6) = 72 - 18 = 54$
$f(-x) = 2(-x)^2 - 3(-x) = 2x^2 + 3x$

99. Draw the numbers in $\{1, 2, 3, ..., 999\}$ that can be written in the form

$$(2n + 1) \cdot 2^{2m}$$

for $n, m \geq 0$. This means that you will have drawn 666 numbers. Then the 667th number you will draw has the form

$$(2n + 1) \cdot 2^{2m+1}$$

which must be twice one of the first 666 numbers.

For Thought

1. True, since $A = P^2/16$.

2. False, rather $y = (x - 1)^2 = x^2 - 2x + 1$.

3. False, rather $(f \circ g)(x) = \sqrt{x - 2}$.

4. True

5. False, since $(h \circ g)(x) = x^2 - 9$.

6. False, rather $f^{-1}(x) = \dfrac{1}{2}x - \dfrac{1}{2}$.

7. True

8. False; g^{-1} does not exist since the graph of g which is a parabola fails the horizontal line test.

9. False, since $y = x^2$ is a function which does not have an inverse function.

10. True

P.4 Exercises

1. composition

3. invertible

5. switch-and-solve

7. $y = 2(3x + 1) - 3 = 6x - 1$

9. $y = (x^2 + 6x + 9) - 2 = x^2 + 6x + 7$

11. $y = 3 \cdot \dfrac{x+1}{3} - 1 = x + 1 - 1 = x$

13. $y = 2 \cdot \dfrac{x+1}{2} - 1 = x + 1 - 1 = x$

15. $f(2) = 5$

17. $f(2) = 5$

19. $f(17) = 3(17) - 1 = 50$

21. $3(x^2 + 1) - 1 = 3x^2 + 2$

23. $(3x - 1)^2 + 1 = 9x^2 - 6x + 2$

25. $\dfrac{x^2+2}{3}$

27. $F = g \circ h$

29. $H = h \circ g$

31. $N = f \circ h$

33. Interchange x and y then solve for y.

$$\begin{aligned}
x &= 3y - 7 \\
\frac{x+7}{3} &= y \\
\frac{x+7}{3} &= f^{-1}(x)
\end{aligned}$$

35. Interchange x and y then solve for y.

$$\begin{aligned}
x &= 2 + \sqrt{y-3} \quad \text{for } x \ge 2 \\
(x-2)^2 &= y - 3 \quad \text{for } x \ge 2 \\
f^{-1}(x) &= (x-2)^2 + 3 \quad \text{for } x \ge 2
\end{aligned}$$

37. Interchange x and y then solve for y.

$$\begin{aligned}
x &= -y - 9 \\
y &= -x - 9 \\
f^{-1}(x) &= -x - 9
\end{aligned}$$

39. Interchange x and y then solve for y.

$$\begin{aligned}
x &= -\frac{1}{y} \\
xy &= -1 \\
f^{-1}(x) &= -\frac{1}{x}
\end{aligned}$$

41. Interchange x and y then solve for y.

$$\begin{aligned}
x &= \sqrt[3]{y-9} + 5 \\
x - 5 &= \sqrt[3]{y-9} \\
(x-5)^3 &= y - 9 \\
f^{-1}(x) &= (x-5)^3 + 9
\end{aligned}$$

43. Interchange x and y then solve for y.

$$\begin{aligned}
x &= (y-2)^2 \quad x \ge 0 \\
\sqrt{x} &= y - 2 \\
f^{-1}(x) &= \sqrt{x} + 2
\end{aligned}$$

45. $(f^{-1} \circ f)(x) = \dfrac{1}{2}(2x-1) + \dfrac{1}{2} = x$ and
$(f \circ f^{-1})(x) = 2\left(\dfrac{1}{2}x + \dfrac{1}{2}\right) - 1 = x$

47. $(f^{-1} \circ f)(x) = 0.25(4x+4) - 1 = x$ and
$(f \circ f^{-1})(x) = 4(0.25x - 1) + 4 = x$

49. We obtain

$$\begin{aligned}
(f^{-1} \circ f)(x) &= \frac{4 - \left(\sqrt[3]{4-3x}\right)^3}{3} \\
&= \frac{4 - (4-3x)}{3} \\
&= \frac{3x}{3} \\
(f^{-1} \circ f)(x) &= x
\end{aligned}$$

and

$$\begin{aligned}
(f \circ f^{-1})(x) &= \sqrt[3]{4 - 3\left(\frac{4-x^3}{3}\right)} \\
&= \sqrt[3]{4 - (4 - x^3)} \\
&= \sqrt[3]{x^3} \\
(f \circ f^{-1})(x) &= x.
\end{aligned}$$

51. We obtain

$$\begin{aligned}
(f^{-1} \circ f)(x) &= \sqrt[5]{\left(\sqrt[3]{x^5 - 1}\right)^3 + 1} \\
&= \sqrt[5]{x^5 - 1 + 1} \\
&= \sqrt[5]{x^5} \\
(f^{-1} \circ f)(x) &= x
\end{aligned}$$

and

$$\begin{aligned}
(f \circ f^{-1})(x) &= \sqrt[3]{\left(\sqrt[5]{x^3 + 1}\right)^5 - 1} \\
&= \sqrt[3]{x^3 + 1 - 1} \\
&= \sqrt[3]{x^3} \\
(f \circ f^{-1})(x) &= x.
\end{aligned}$$

53. $f^{-1}(x) = \dfrac{x-2}{3}$

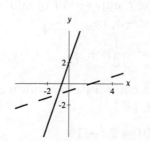

55. $f^{-1}(x) = \sqrt{x+4}$

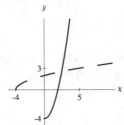

57. $f^{-1}(x) = \sqrt[3]{x}$

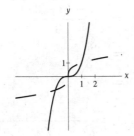

59. $f^{-1}(x) = (x+3)^2$ for $x \geq -3$

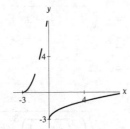

61. The inverse is the composition of subtracting 2 and dividing by 5, i.e., $f^{-1}(x) = \dfrac{x-2}{5}$

63. The inverse is the composition of adding 88 and dividing by 2, i.e.,
$$f^{-1}(x) = \frac{x+88}{2} = \frac{1}{2}x + 44$$

65. The inverse is the composition of subtracting 7 and dividing by -3, i.e.,
$$f^{-1}(x) = \frac{x-7}{-3} = -\frac{1}{3}x + \frac{7}{3}$$

67. The inverse is the composition of subtracting 4 and dividing by -3, i.e.,
$$f^{-1}(x) = \frac{x-4}{-3} = -\frac{1}{3}x + \frac{4}{3}$$

69. The inverse is the composition of adding 9 and multiplying by 2, i.e.,
$$f^{-1}(x) = 2(x+9) = 2x + 18$$

71. The inverse is the composition of subtracting 3 and taking the multiplicative inverse, i.e.,
$$f^{-1}(x) = \frac{1}{x-3}$$

73. The inverse is the composition of adding 9 and raising an expression to the third power, i.e.,
$$f^{-1}(x) = (x+9)^3.$$

75. The inverse is the composition of adding 7, dividing by 2, and taking the cube root, i.e.,
$$f^{-1}(x) = \sqrt[3]{\frac{x+7}{2}}.$$

77. $A = d^2/2$

79. Let x be the number of ice cream bars sold. Then $W(x) = (0.20)(0.40x+200) = 0.08x+40$.

81. $C = 1.08P$ expresses the total cost as a function of the purchase price; and $P = C/1.08$ is the purchase price as a function of the total cost.

83.
$$r = \sqrt{5.625 \times 10^{-5} - \frac{V}{500}}$$
where $0 \leq V \leq 0.028125$

85. When $V = \$18,000$, the depreciation rate is
$$1 - \left(\frac{18,000}{50,000}\right)^{1/5} \approx 0.1848$$
or the depreciation rate is 18.48%.
Solving for V, we obtain
$$\left(\frac{V}{50,000}\right)^{1/5} = 1-r$$
$$\frac{V}{50,000} = (1-r)^5$$
$$V = 50,000(1-r)^5.$$

87. Since $g^{-1}(x) = \dfrac{x+5}{3}$ and $f^{-1}(x) = \dfrac{x-1}{2}$, we have
$$g^{-1} \circ f^{-1}(x) = \frac{\dfrac{x-1}{2}+5}{3} = \frac{x+9}{6}.$$

Likewise, since $(f \circ g)(x) = 6x - 9$, we get

$$(f \circ g)^{-1}(x) = \frac{x + 9}{6}.$$

Hence, $(f \circ g)^{-1} = g^{-1} \circ f^{-1}$.

89. $x^2 + (y - 1)^2 = 9$

91. function

93. Domain $(-\infty, \infty)$, range $[1, \infty)$

95. Note, if $a > b > 0$ and $n > 0$, then

$$\frac{a}{n} + \frac{b}{n+1} > \frac{b}{n} + \frac{a}{n+1}.$$

Thus, the arrangement with the largest sum is

$$\frac{2015}{1} + \frac{2014}{2} + \ldots + \frac{2}{2014} + \frac{1}{2015}.$$

Chapter P Review Exercises

1. $\sqrt{49 \cdot 2} = 7\sqrt{2}$

3. $\dfrac{3}{\sqrt{5}} \cdot \dfrac{\sqrt{5}}{\sqrt{5}} = \dfrac{3\sqrt{5}}{5}$

5. $\dfrac{\sqrt{5}}{\sqrt{6}} \cdot \dfrac{\sqrt{6}}{\sqrt{6}} = \dfrac{\sqrt{30}}{6}$

7. $\dfrac{8}{\sqrt{2}} \cdot \dfrac{\sqrt{2}}{\sqrt{2}} = \dfrac{8\sqrt{2}}{2} = 4\sqrt{2}$

9. The distance is $\sqrt{(-3 - 2)^2 + (5 - (-6))^2} =$
$\sqrt{(-5)^2 + 11^2} = \sqrt{25 + 121} = \sqrt{146}$.

The midpoint is $\left(\dfrac{-3 + 2}{2}, \dfrac{5 - 6}{2} \right) =$

$\left(-\dfrac{1}{2}, -\dfrac{1}{2} \right)$.

11. The distance is $\sqrt{\left(\pi - \dfrac{\pi}{2} \right)^2 + (1 - 1)^2} =$

$\sqrt{\left(\dfrac{\pi}{2} \right)^2 + 0} = \dfrac{\pi}{2}$. The midpoint is

$\left(\dfrac{\pi/2 + \pi}{2}, \dfrac{1 + 1}{2} \right) = \left(\dfrac{3\pi}{4}, 1 \right)$.

13. Circle with center at $(0, 0)$ and radius 3

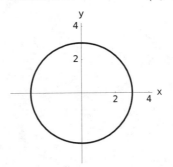

15. Circle with center at $(0, 1)$ and radius 1

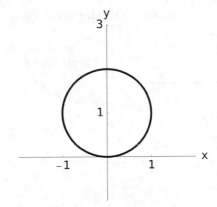

17. The line through the points $(10, 0)$ and $(0, -4)$.

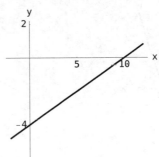

19. The vertical line through the point $(5, 0)$.

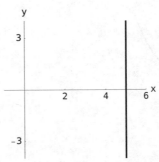

21. Domain $(-\infty, \infty)$ and range $(-\infty, \infty)$

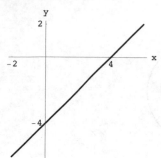

23. Domain $(-\infty, \infty)$ and range $\{4\}$.

     The horizontal line through the point $(0, 4)$.

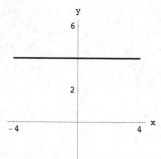

25. Domain $(-\infty, \infty)$ and range $[-3, \infty)$

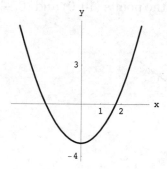

27. Domain $(-\infty, \infty)$ and range $(-\infty, \infty)$

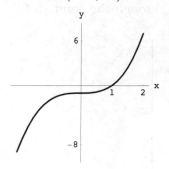

29. Domain $[1, \infty)$ and range $[0, \infty)$

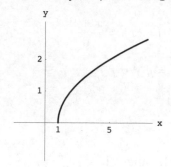

31. Domain $(-\infty, \infty)$ and range $[-4, \infty)$

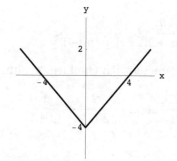

33. $9 + 3 = 12$

35. $24 - 7 = 17$

37. $g(12) = 17$

39. $f(-3) = 12$

41.

$$
\begin{aligned}
f(g(x)) &= f(2x - 7) \\
&= (2x - 7)^2 + 3 \\
&= 4x^2 - 28x + 52
\end{aligned}
$$

43. $g\left(\dfrac{x + 7}{2}\right) = (x + 7) - 7 = x$

45. $g^{-1}(x) = \dfrac{x + 7}{2}$

47. $f(x) = \sqrt{x}$, $g(x) = 2\sqrt{x + 3}$; left by 3, stretch by 2

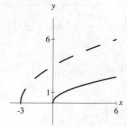

49. $f(x) = |x|$, $g(x) = -2|x + 2| + 4$; left by 2, stretch by 2, reflect about x-axis, up by 4

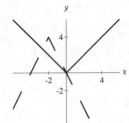

51. $f(x) = x^2$, $g(x) = (x - 2)^2 + 1$; right by 2 and up by 1

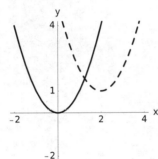

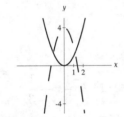

53. $N = h \circ j$

55. $R = h \circ g$

57. $F = f \circ g$

59. $H = f \circ g \circ j$

61. $y = |x| - 3$, domain is $(-\infty, \infty)$, range is $[-3, \infty)$

63. $y = -2|x| + 4$, domain is $(-\infty, \infty)$, range is $(-\infty, 4]$

range is $(-\infty, 2]$

65. $y = |x + 2| + 1$, domain is $(-\infty, \infty)$, range is $[1, \infty)$

67. Symmetry: y-axis

69. Symmetric about the origin

71. Neither symmetry

73. Symmetric about the y-axis

75. inverse functions,
$$f(x) = \sqrt{x + 3}, g(x) = x^2 - 3 \text{ for } x \geq 0$$

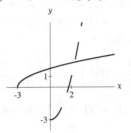

77. inverse functions,
$$f(x) = 2x - 4, g(x) = \frac{1}{2}x + 2$$

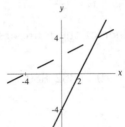

79. $f^{-1}(x) = \dfrac{x}{5}$

81. $f^{-1}(x) = \dfrac{x + 21}{3}$ or $f^{-1}(x) = \dfrac{1}{3}x + 7$

83. $f^{-1}(x) = \sqrt[3]{x + 1}$

85. $f^{-1}(x) = \dfrac{1}{x} + 3$

87. $f^{-1}(x) = \sqrt{x}$

89. $\sqrt{(\sqrt{5})^2 + 2^2} = \sqrt{9} = 3$

91. From $3x - 4(0) = 9$, we get that the x-intercept is $(3, 0)$. From $3(0) - 4y = 9$, we find that the y-intercept is $\left(0, -\dfrac{9}{4}\right)$.

93. $(x + 3)^2 + (y - 5)^2 = 3$

95. The given circle is $(x - 3)^2 + y^2 = 1$. Let $P(a, \pm\sqrt{1 - (a - 3)^2})$ be the points of tangency for the tangent line that passes through the origin.

Let $A^2 = a^2 + 1 - (a - 3)^2$ be the square of the distance between P and the origin. Simplifying, $A = 6a - 8$.

Let $B^2 = (a-3)^2 + 1 - (a-3)^2$ be the square of the distance between P and $(3,0)$. Simplifying, $B = 1$.

By the Pythagorean theorem, $A^2 + B^2 = 9$. Equivalently, $6a - 8 + 1 = 9$ or $a = \frac{8}{3}$. Then point P has coordinates

$$P\left(\frac{8}{3}, \pm\sqrt{1 - (\tfrac{8}{3} - 3)^2}\right) = P\left(\frac{8}{3}, \pm\frac{2\sqrt{2}}{3}\right).$$

Thus, the tangent lines through the origin and P are given by

$$y = \pm\frac{\sqrt{2}}{4}x.$$

97. Since $C = 2\pi r$, we get $r = \dfrac{C}{2\pi}$.

99. No, since the product of the numbers from 1 through 9 is

$$1 \cdot 2 \cdot 3 \cdots 9 = 362{,}880$$

and $71^3 = 357911$.

Chapter P Test

1. Domain $(-\infty, \infty)$, range $[0, \infty)$

2. Domain $[9, \infty)$, range $[0, \infty)$

3. Domain $(-\infty, \infty)$, range $(-\infty, 2]$

4. The graph of $y = 2x - 3$ includes the points $(3/2, 0), (0, -3)$

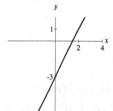

5. The graph of $f(x) = \sqrt{x} - 5$ goes through $(5, 0)$, $(9, 2)$.

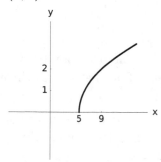

6. The graph of $y = 2|x| - 4$ includes the points $(0, -4), (\pm 3, 2)$

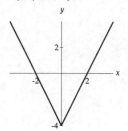

7. The graph of $y = -(x-2)^2 + 5$ is a parabola with vertex $(2, 5)$

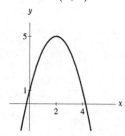

8. $\sqrt{9} = 3$

9. $3(2) - 1 = 5$

10. $f(5) = \sqrt{7}$

11. $g^{-1}(x) = \dfrac{x+1}{3}$ or $g^{-1}(x) = \dfrac{1}{3}x + \dfrac{1}{3}$

12. Using the answer from Exercise 11, we find $g^{-1}(20) = \dfrac{20+1}{3} = 7$.

13. $\sqrt{72} = 6\sqrt{2}$

14. $\dfrac{\sqrt{25}}{\sqrt{6}} = \dfrac{5\sqrt{6}}{6}$

15. $\dfrac{20}{2\sqrt{2}} = 5\sqrt{2}$

16. $\left(\dfrac{3 + (-3)}{2}, \dfrac{5 + 6}{2}\right) = \left(0, \dfrac{11}{2}\right)$

17. $\sqrt{(-2-3)^2 + (4 - (-1))^2} = \sqrt{25 + 25} = 5\sqrt{2}$.

18. $(x+4)^2 + (y-1)^2 = 7$

19. Center $(4, -1)$, radius $\sqrt{6}$

20. Solving $3a - 9 = 6$, we obtain $a = 5$.

21. Symmetric about the y-axis
since $f(-x) = f(x)$

22. Interchange x and y then solve for y.

$$\begin{aligned} x &= 3 + \sqrt[3]{y - 2} \\ x - 3 &= \sqrt[3]{y - 2} \\ (x - 3)^3 &= y - 2 \\ (x - 3)^3 + 2 &= y \end{aligned}$$

The inverse is $g^{-1}(x) = (x - 3)^3 + 2$.

23. Since $A = \left(\dfrac{p}{4}\right)^2$, we get $A = \dfrac{p^2}{16}$.

For Thought

1. True   **2.**   True

3. True, since one complete revolution is 360°.

4. False, the number of degrees in the intercepted arc of a circle is the degree measure.

5. False, the degree measure is negative if the rotation is clockwise.

6. True, the terminal side of 540° lies in the negative x-axis.

7. True

8. False, since $-365°$ lies in quadrant IV while 5° lies in quadrant I.

9. True, since $25°60' + 6' = 26°6'$.

10. False, since $25 + 20/60 + 40/3600 = 25.3444....$

1.1 Exercises

1. angle

3. standard position

5. obtuse

7. coterminal

9. minute

11. Substitute $k = 1, 2, -1, -2$ into $60° + k \cdot 360°$.
Coterminal angles are
$420°, 780°, -300°, -660°$.
There are other coterminal angles.

13. Substitute $k = 1, 2, -1, -2$ into $30° + k \cdot 360°$.
Coterminal angles are
$390°, 750°, -330°, -690°$.
There are other coterminal angles.

15. Substitute $k = 1, 2, -1, -2$ into $225° + k \cdot 360°$.
Coterminal angles are
$585°, 945°, -135°, -495°$.
There are other coterminal angles.

17. Substitute $k = 1, 2, -1, -2$ into $-45° + k \cdot 360°$.
Coterminal angles are
$315°, 675°, -405°, -765°$.
There are other coterminal angles.

19. Substitute $k = 1, 2, -1, -2$ into $-90° + k \cdot 360°$.
Coterminal angles are
$270°, 630°, -450°, -810°$.
There are other coterminal angles.

21. Substitute $k = 1, 2, -1, -2$ into
$-210° + k \cdot 360°$. Coterminal angles are
$150°, 510°, -570°, -930°$.
There are other coterminal angles.

23. Yes, since $40° - (-320°) = 360°$

25. No, since $4° - (-364°) = 368° \neq k \cdot 360°$
for any integer k.

27. Yes, since $1235° - 155° = 3 \cdot 360°$.

29. Yes, since $22° - (-1058)° = 3 \cdot 360°$.

31. No, since $312.4° - (-227.6°) = 540° \neq k \cdot 360°$
for any integer k.

33. Quadrant I

35. $-125°$ lies in Quadrant III since
$-125° + 360° = 235°$ and $180° < 235° < 270°$

37. -740 lies in Quadrant IV since
$-740° + 2 \cdot 360° = -20°$ and $-20°$ lies
in Quadrant IV.

39. $933°$ lies in Quadrant III since
$933° - 2 \cdot 360° = 213°$ and $213°$ lies
in Quadrant III.

41. $-310°$, since $50° - 360° = -310°$

43. $-220°$, since $140° - 360° = -220°$

45. $-90°$, since $270° - 360° = -90°$

47. $40°$, since $400° - 360° = 40°$

49. $20°$, since $-340° + 360° = 20°$

51. $340°$, since $-1100° + 4 \cdot 360° = 340°$

53. $180.54°$, since $900.54° - 2 \cdot 360° = 180.54°$

55. c   **57.**   e

59. h   **61.**   g

63. $13° + \dfrac{12°}{60} = 13.2°$

65. $-8° - \dfrac{30°}{60} - \dfrac{18}{3600}° = -8.505°$

67. $28° + \dfrac{5}{60}° + \dfrac{9}{3600}° \approx 28.0858°$

69. $155° + \dfrac{34°}{60} + \dfrac{52}{3600}° \approx 155.5811°$

71. $75.5° = 75°30'$ since $0.5(60) = 30$

73. $39.4° = 39°24'$ since $0.4(60) = 24$

75. $-17.33° = -17°19'48''$ since $0.33(60) = 19.8$ and $0.8(60) = 48$

77. $18.123° \approx 18°7'23''$ since $0.123(60) = 7.38$ and $0.38(60) \approx 23$

79. $24°15' + 33°51' = 57°66' = 58°6'$

81. $55°11' - 23°37' = 54°71' - 23°37' = 31°34'$

83. $16°23'41'' + 44°43'39'' = 60°66'80'' = 60°67'20'' = 61°7'20''$

85. $90° - 7°44'35'' = 89°59'60'' - 7°44'35'' = 82°15'25''$

87. $66°43'6'' - 5°51'53'' = 65°102'66'' - 5°51'53'' = 60°51'13''$

$33°98'72'' - 9°49'18'' = 24°49'54''$

89. $2(32°36'37'') = 64°72'74'' = 64°73'14'' = 65°13'14''$

91. $3(15°53'42'') = 45°159'126'' = 45°161'6'' = 47°41'6''$

93. $(43°13'8'')/2 = (42°73'8'')/2 = (42°72'68'')/2 = 21°36'34''$

95. $(13°10'9'')/3 = (12°70'9'')/3 = (12°69'69'')/3 = 4°23'23''$

97. $\alpha = 180° - 88°40' - 37°52' = 180° - 126°32' = 179°60' - 126°32' = 53°28'$

99. $\alpha = 180° - 90° - 48°9'6'' = 89°59'60'' - 48°9'6'' = 41°50'54''$

101. $\alpha = 180° - 140°19'16'' = 179°59'60'' - 140°19'16'' = 39°40'44''$

103. $\alpha = 90° - 75°5'6'' = 89°59'60'' - 75°5'6'' = 14°54'54''$

105. Since $0.17647(60) = 10.5882$ and $0.5882(60) = 35.292$, we find $21.17647° \approx 21°10'35.3''$.

107. Since $73°37' \approx 73.6167°$, $49°41' \approx 49.6833°$, and $56°42' = 56.7000°$, the sum of the numbers in decimal format is $180°$. Also,

$$73°37' + 49°41' + 56°42' = 178°120' = 180°.$$

109. We find $108°24'16'' \approx 108.4044°$, $68°40'40'' \approx 68.6778°$, $84°42'51'' \approx 84.7142°$, and $98°12'13'' \approx 98.2036°$. The sum of these four numbers in decimal format is $360°$. Also, $108°24'16'' + 68°40'40'' + 84°42'51'' + 98°12'13'' = 360°$.

111. At 3:20, the hour hand is at an angle

$$\frac{20}{60} \times 30° = 10°$$

below the 3 in a standard clock.

Since the minute hand is at 4 and there are $30°$ between 3 and 4, the angle between the hour and minute hands is

$$30° - 10° = 20°.$$

113. $f(g(2)) = f(4) = 15$ and $g(f(2)) = g(3) = 7$

115. $\sqrt{(4-1)^2 + (5-1)^2} = \sqrt{9 + 16} = 5$

117. $\sqrt{5^2 + 5^2} = 5\sqrt{2}$ feet

119. Let r be the radius of the circle. Let p be the distance from the lower left most corner of the 1-by-1 square to the point of tangency of the left most, lower most circle at the base of the 1-by-1 square. By the Pythagorean Theorem,

$$(p + r)^2 + (r + 1 - p)^2 = 1$$

or equivalently

$$p^2 + r^2 + r - p = 0. \tag{1}$$

Since the area of the four triangles plus the area of the small square in the middle is 1, we obtain

$$4r^2 + 4\left[\frac{1}{2}(p+r)(r+1-p)\right] = 1$$

or

$$6r^2 - 2p^2 + 2p + 2r = 1. \qquad (2)$$

Multiply (1) by two and add the result to (2). We obtain

$$8r^2 + 4r = 1.$$

The solution is $r = (\sqrt{3} - 1)/4$.

For Thought

1. False, in a negative angle the rotation is clockwise.

2. False, the radius is 1.

3. True, since the circumeference is $2\pi r$ where r is the radius.

4. True, since $s = \alpha r$ 5. True

6. False, one must multiply by $\frac{\pi}{180}$.

7. True   8. False, rather $45° = \frac{\pi}{4}$ rad.

9. True

10. True, rather the length of arc is
$$s = \alpha \cdot r = \frac{\pi}{4} \cdot 4 = 1.$$

1.2 Exercises

1. unit

3. $s = \alpha r$

5. $30° = \dfrac{\pi}{6}$, $45° = \dfrac{\pi}{4}$, $60° = \dfrac{\pi}{3}$, $90° = \dfrac{\pi}{2}$,
$120° = \dfrac{2\pi}{3}$, $135° = \dfrac{3\pi}{4}$, $150° = \dfrac{5\pi}{6}$, $180° = \pi$,
$210° = \dfrac{7\pi}{6}$, $225° = \dfrac{5\pi}{4}$, $240° = \dfrac{4\pi}{3}$,
$270° = \dfrac{3\pi}{2}$, $300° = \dfrac{5\pi}{3}$, $315° = \dfrac{7\pi}{4}$,
$330° = \dfrac{11\pi}{6}$, $360° = 2\pi$

7. $45 \cdot \dfrac{\pi}{180} = \dfrac{\pi}{4}$

9. $90 \cdot \dfrac{\pi}{180} = \dfrac{\pi}{2}$

11. $120 \cdot \dfrac{\pi}{180} = \dfrac{2\pi}{3}$

13. $150 \cdot \dfrac{\pi}{180} = \dfrac{5\pi}{6}$

15. $18 \cdot \dfrac{\pi}{180} = \dfrac{\pi}{10}$

17. $\dfrac{\pi}{3} \cdot \dfrac{180}{\pi} = 60°$

19. $\dfrac{5\pi}{12} \cdot \dfrac{180}{\pi} = 75°$

21. $\dfrac{3\pi}{4} \cdot \dfrac{180}{\pi} = 135°$

23. $-6\pi \cdot \dfrac{180}{\pi} = -1080°$

25. $2.39 \cdot \dfrac{180}{\pi} \approx 136.937°$

27. $-0.128 \cdot \dfrac{180}{\pi} \approx -7.334°$

29. $37.4\left(\dfrac{\pi}{180}\right) \approx 0.653$

31. $\left(-13 - \dfrac{47}{60}\right) \cdot \dfrac{\pi}{180} \approx -0.241$

33. $\left(53 + \dfrac{37}{60} + \dfrac{6}{3600}\right) \cdot \dfrac{\pi}{180} \approx 0.936$

35. Substitute $k = 1, 2, -1, -2$ into $\dfrac{\pi}{3} + k \cdot 2\pi$,
coterminal angles are $\dfrac{7\pi}{3}, \dfrac{13\pi}{3}, -\dfrac{5\pi}{3}, -\dfrac{11\pi}{3}$.
There are other coterminal angles.

37. Substitute $k = 1, 2, -1, -2$ into $\dfrac{\pi}{2} + k \cdot 2\pi$,
coterminal angles $\dfrac{5\pi}{2}, \dfrac{9\pi}{2}, -\dfrac{3\pi}{2}, -\dfrac{7\pi}{2}$.
There are other coterminal angles.

39. Substitute $k = 1, 2, -1, -2$ into $\dfrac{2\pi}{3} + k \cdot 2\pi$,
coterminal angles are $\dfrac{8\pi}{3}, \dfrac{14\pi}{3}, -\dfrac{4\pi}{3}, -\dfrac{10\pi}{3}$.
There are other coterminal angles.

41. Substitute $k = 1, 2, -1, -2$ into $1.2 + k \cdot 2\pi$, coterminal angles are about 7.5, 13.8, -5.1, -11.4. There are other coterminal angles.

43. Quadrant I

45. Quadrant III

47. $-\dfrac{13\pi}{8}$ lies in Quadrant I since

$$-\frac{13\pi}{8} + 2\pi = \frac{3\pi}{8}$$

49. $\dfrac{17\pi}{3}$ lies in Quadrant IV since

$$\frac{17\pi}{3} - 4\pi = \frac{5\pi}{3}$$

51. 3 lies in Quadrant II since $\dfrac{\pi}{2} < 3 < \pi$

53. -7.3 lies in Quadrant IV since $-7.3 + 4\pi \approx 5.3$ and 5.3 lies in Quadrant IV

55. g **57.** b

59. h **61.** d

63. π, since $3\pi - 2\pi = \pi$ and π is the smallest positive coterminal angle

65. $\dfrac{3\pi}{2}$, since $-\dfrac{\pi}{2} + 2\pi = \dfrac{3\pi}{2}$

67. $\dfrac{9\pi}{2} - 4\pi = \dfrac{\pi}{2}$

69. $-\dfrac{5\pi}{3} + 2\pi = \dfrac{\pi}{3}$

71. $-\dfrac{13\pi}{3} + 6\pi = \dfrac{5\pi}{3}$

73. $8.32 - 2\pi \approx 2.04$

75. $\dfrac{4\pi}{4} - \dfrac{\pi}{4} = \dfrac{3\pi}{4}$

77. $\dfrac{3\pi}{6} + \dfrac{2\pi}{6} = \dfrac{5\pi}{6}$

79. $\dfrac{2\pi}{4} + \dfrac{\pi}{4} = \dfrac{3\pi}{4}$

81. $\dfrac{\pi}{3} - \dfrac{3\pi}{3} = -\dfrac{2\pi}{3}$

83. $s = 12 \cdot \dfrac{\pi}{4} = 3\pi \approx 9.4$ ft

85. $s = 4000 \cdot \dfrac{3\pi}{180} \approx 209.4$ miles

87. $s = (26.1)(1.3) \approx 33.9$ m

89. radius is $r = \dfrac{s}{\alpha} = \dfrac{1}{1} = 1$ mile.

91. radius is $r = \dfrac{s}{\alpha} = \dfrac{10}{\pi} \approx 3.2$ km

93. radius is $r = \dfrac{s}{\alpha} = \dfrac{500}{\pi/6} \approx 954.9$ ft

95. $A = \dfrac{\alpha r^2}{2} = \dfrac{(\pi/6)6^2}{2} = 3\pi$

97. $A = \dfrac{\alpha r^2}{2} = \dfrac{(\pi/3)12^2}{2} = 24\pi$

99. Distance from Peshtigo to the North Pole is

$$s = r\alpha = 3950 \left(45 \cdot \frac{\pi}{180} \right) \approx 3102 \text{ miles}$$

101. The central angle is $\alpha = \dfrac{2000}{3950} \approx$

0.506329 radians $\approx 0.506329 \cdot \dfrac{180}{\pi} \approx 29.0°$

103. Since $7° \approx 0.12217305$, the radius of the earth according to Eratosthenes is

$$r = \frac{s}{\alpha} \approx \frac{800}{0.12217305} \approx 6548.089 \text{ km}.$$

Thus, using Eratosthenes' radius, the circumference is

$$2\pi r \approx 41,143 \text{ km}.$$

Using $r = 6378$ km, circumference is 40,074 km.

105. The length of the arc intercepted by the central angle $\alpha = 20°$ in a circle of radius 6 in. is

$$s = 6 \left(20 \cdot \frac{\pi}{180} \right) = \frac{2\pi}{3}.$$

Since the radius of the cog is 2 in., the cog rotates through an angle

$$\alpha = \frac{s}{r} = \frac{2\pi/3}{2} = \frac{\pi}{3} \text{ radians} = 60°.$$

107. If s is the arc length, the radius is $r = \frac{6-s}{2}$. Let θ be the central angle such that

$$s = r\theta = \frac{(6-s)\theta}{2}.$$

Solving for s, we find

$$s = \frac{6\theta}{\theta+2}.$$

The area of the sector is

$$A = \frac{\theta r^2}{2} = \frac{\theta}{2}\left(\frac{s}{\theta}\right)^2 = \frac{s^2}{2\theta}.$$

Substituting, we obtain

$$A = \frac{1}{2\theta}\left(\frac{6\theta}{\theta+2}\right)^2$$
$$= \frac{18\theta}{(\theta+2)^2}.$$

109. Note, the fraction of the area of the circle of radius r intercepted by the central angle $\frac{\pi}{6}$ is $\frac{1}{12} \cdot \pi r^2$. So, the area watered in one hour is

$$\frac{1}{12} \cdot \pi 150^2 \approx 5890 \text{ ft}^2.$$

111. Note, the radius of the pizza is 8 in. Then the area of each of the six slices is

$$\frac{1}{6} \cdot \pi 8^2 \approx 33.5 \text{ in.}^2$$

113. A region S bounded by a chord and a circle of radius 10 meters is shaded below. The central angle is $60°$. The area A_s of S may be obtained by subtracting the area of an equilateral triangle from the area of a sector. That is,

$$A_s = 100\left(\frac{\pi}{6} - \frac{\sqrt{3}}{4}\right).$$

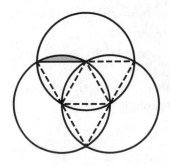

The common region bounded by two or three circles consists of 4 equilateral triangles and six regions like S.

The area of the region inside the higher circle but outside the common region is $100(\pi/2 - 2A_s)$. There are two other such regions for the two circles on the left and right.

Then the total area A watered by the three circular sprinklers is

$$A = \left[100\sqrt{3} + 6A_s + 3\left(\frac{100\pi}{2} - 2A_s\right)\right]$$
$$= 100\left[\frac{3\pi}{2} + \sqrt{3}\right] \text{ m}^2$$
$$\approx 644.4 \text{ m}^2.$$

115. a) Given angle α (in degrees) as in the problem, the radius r of the cone must satisfy $2\pi r = 8\pi - 4\alpha\frac{\pi}{180}$; note, 8π in. is the circumference of a circle with radius 4 in. Then $r = 4 - \frac{\alpha}{90}$. Note, $h = \sqrt{16 - r^2}$ by the Pythagorean theorem. Since the volume $V(\alpha)$ of the cone is $\frac{\pi}{3}r^2 h$, we find

$$V(\alpha) = \frac{\pi}{3}\left(4 - \frac{\alpha}{90}\right)^2 \sqrt{16 - \left(4 - \frac{\alpha}{90}\right)^2}$$

This reduces to

$$V(\alpha) = \frac{\pi(360 - \alpha)^2\sqrt{720\alpha - \alpha^2}}{2,187,000}$$

If $\alpha = 30°$, then $V(30°) \approx 22.5 \text{ inches}^3$.

b) As shown in part a), the volume of the cone obtained by an overlapping angle α is

$$V(\alpha) = \frac{\pi(360 - \alpha)^2\sqrt{720\alpha - \alpha^2}}{2,187,000}$$

119. angle, rays

121. $x = \pi$

123. right

125. a) The contestants that leave the table are # 1, # 3, # 5, # 7, # 9, # 11, # 13, # 4, # 8, # 12, # 2, #6 (in this order). Thus, contestant # 10 is the unlucky contestant.

b) Let $n = 8$. The contestants that leave the table are # 1, # 3, # 5, # 7, # 2, # 6, # 4 (in this order). Thus, contestant # 8 is the unlucky contestant.

Let $n = 16$. The contestants that leave the table are # 1, # 3, # 5, # 7, # 9, # 11, # 13, # 15, # 2, # 6, # 10, # 14, # 4, # 12, # 8 (in this order). Thus, contestant # 16 is the unlucky contestant.

Let $n = 41$. The contestants that leave the table are # 1, # 3, # 5, # 7, # 9, # 11, # 13, # 15, # 17, # 19, # 21, # 23, # 25, # 27, # 29, # 31, # 33, # 35, # 37, # 39, # 4, # 8 # 12, # 16, # 20, # 24, # 28, # 32, # 36, # 40, # 6, # 14, # 22, # 30, # 38, # 10, # 26, # 2, # 34 (in this order). Thus, contestant # 18 is the unlucky contestant.

c) Let $m > 1$ and let n satisfy

$$2^m < n \leq 2^{m+1} \qquad (3)$$

where

$$n = 2^m + k. \qquad (4)$$

We claim the unlucky number is $2k$. One can check the claim is true for all n when $m = 1$. Suppose the claim is true for $m - 1$ and all such n.

Consider the case when k is an even integer satisfying (3) and (4). After selecting the survivors in round 1, the remaining contestants are

$$2, 4, 6, ..., 2^m + k.$$

Renumber, the above contestants by the rule

$$f(x) = x/2$$

so that the remaining contestants are renumbered as

$$1, 2, 3, ..., 2^{m-1} + k/2.$$

Note,

$$2^{m-1} < 2^{m-1} + k/2 \leq 2^m$$

Since the claim is true for $m - 1$, the unlucky contestant in the renumbering is

$$2\left(\frac{k}{2}\right) = k.$$

But by the renumbering, we find

$$f(x) = \frac{x}{2} = k$$

or the unlucky contestant is $2k$. In particular, if $n = 2^m$ then the unlucky contestant is

$$2k = 2 \cdot 2^{m-1} = 2^m.$$

Finally, let k be an odd integer satisfying (3) and (4). After selecting the survivors in round 1, the remaining contestants are

$$2, 4, 6, ..., 2^m + (k - 1).$$

Note, the next survivor is 4 since the last survivor chosen is $2^m + k$.

Renumber, the above contestants by the rule

$$g(x) = \frac{x}{2} - 1$$

so the remaining contestant are renumbered as follows:

$$0, 1, 2, 3, ..., [2^{m-1} + (k - 1)/2 - 1].$$

Since the claim is true for $m - 1$, the unlucky contestant using the previous renumbering is

$$2\left(\frac{k-1}{2}\right) \quad \text{or} \quad k - 1$$

Using the original numbering, the unlucky contestant is

$$g(x) = \frac{x}{2} - 1 = k - 1$$

or

$$x = 2k = 2(n - 2^m).$$

For Thought

1. False, $\dfrac{240 \text{ rev}}{\text{min}} = \dfrac{(240)(6) \text{ rev}}{\text{hr}}$.

2. True

3. False, since $\dfrac{4 \text{ rev}}{\text{sec}} \cdot \dfrac{2\pi \text{ ft}}{1 \text{ rev}} = \dfrac{8\pi \text{ ft}}{\text{sec}}$.

4. False, since $\dfrac{5\pi \text{ rad}}{1 \text{ hr}} \cdot \dfrac{60 \text{ min}}{1 \text{ hr}} = \dfrac{300\pi \text{ rad} \cdot \text{min}}{\text{hr}^2}$.

5. False, it is angular velocity.

6. False, it is linear velocity.

7. False, since 40 inches/second is linear velocity.

8. True, since 1 rev/sec is equivalent to $\omega = 2\pi$ radians/sec, we get that the linear velocity is $v = r\omega = 1 \cdot 2\pi = 2\pi$ ft/sec.

9. True

10. False, Miami has a faster linear velocity than Boston. Note, Miami's distance from the axis of the earth is farther than that of Boston's.

1.3 Exercises

1. angular velocity

3. linear velocity

5. $\dfrac{300 \text{ rad}}{60 \text{ min}} = 5$ rad/min

7. $\dfrac{4(2\pi) \text{ rad}}{\text{sec}} = 8\pi$ rad/sec

9. $\dfrac{55(6\pi) \text{ ft}}{\text{min}} \approx 1036.7$ ft/min

11. $\dfrac{10(60) \text{ rev}}{2\pi \text{ hr}} \approx 95.5$ rev/hr

13. $\dfrac{30 \text{ rev}}{\text{min}} \cdot \dfrac{2\pi \text{ rad}}{\text{rev}} = 60\pi \approx 188.5$ rad/min

15. $\dfrac{120 \text{ rev}}{\text{hr}} \cdot \dfrac{1 \text{ hr}}{60 \text{ min}} = 2$ rev/min

17. $\dfrac{180 \text{ rev}}{\text{sec}} \cdot \dfrac{3600 \text{ sec}}{1 \text{ hr}} \cdot \dfrac{2\pi \text{ rad}}{1 \text{ rev}} \approx$
    $4,071,504.1$ rad/hr

19. $\dfrac{30 \text{ mi}}{\text{hr}} \cdot \dfrac{1 \text{ hr}}{3600 \text{ sec}} \cdot \dfrac{5280 \text{ ft}}{1 \text{ mi}} \approx$
    44 ft/sec

21. $\dfrac{500 \text{ rev}}{\text{sec}} \cdot \dfrac{2\pi \text{ rad}}{1 \text{ rev}} = 1000\pi \approx 3141.6$ rad/sec

23. $\dfrac{433.2 \text{ rev}}{\text{min}} \cdot \dfrac{1 \text{ min}}{60 \text{ sec}} \cdot \dfrac{2\pi \text{ rad}}{1 \text{ rev}} \approx$
    45.4 rad/sec

25. $\dfrac{50,000 \text{ rev}}{\text{day}} \cdot \dfrac{1 \text{ day}}{3600(24) \text{ sec}} \cdot \dfrac{2\pi \text{ rad}}{1 \text{ rev}} \approx$
    3.6 rad/sec

27. Convert rev/min to rad/hr:

    $$\dfrac{3450 \text{ rev}}{\text{min}} \cdot \dfrac{60 \text{ min}}{1 \text{ hr}} \cdot \dfrac{2\pi \text{ rad}}{1 \text{ rev}} = \dfrac{120\pi(3450) \text{ rad}}{1 \text{ hr}}.$$

    Since arc length is $s = r\alpha$, linear velocity is

    $$v = \left((3 \text{ in.}) \cdot \dfrac{1 \text{ mi}}{5280(12) \text{ in.}} \right) \cdot \dfrac{120\pi(3450) \text{ rad}}{1 \text{ hr}}$$
    ≈ 61.6 mph.

29. Convert rev/min to rad/hr:

    $$\dfrac{3450 \text{ rev}}{\text{min}} \cdot \dfrac{60 \text{ min}}{1 \text{ hr}} \cdot \dfrac{2\pi \text{ rad}}{1 \text{ rev}} = \dfrac{120\pi(3450) \text{ rad}}{1 \text{ hr}}.$$

    Since arc length is $s = r\alpha$, linear velocity is

    $$v = \left((5 \text{ in.}) \cdot \dfrac{1 \text{ mi}}{5280(12) \text{ in.}} \right) \cdot \dfrac{120\pi(3450) \text{ rad}}{1 \text{ hr}}$$
    ≈ 102.6 mph.

31. Note,

    $$\dfrac{3450 \text{ rev}}{\text{min}} \cdot \dfrac{60 \text{ min}}{1 \text{ hr}} \cdot \dfrac{2\pi \text{ rad}}{1 \text{ rev}} = \dfrac{120\pi(3450) \text{ rad}}{1 \text{ hr}}$$

    and arc length is $s = r\alpha$.
    The linear velocity is

    $$v = \left((7 \text{ in.}) \cdot \dfrac{1 \text{ mi}}{5280(12) \text{ in.}} \right) \cdot \dfrac{120\pi(3450) \text{ rad}}{1 \text{ hr}}$$
    ≈ 143.7 mph

33. The angular velocity is

    $w = 45(2\pi) = 90\pi \approx 282.7$ rad/min.

    Sinc arc length $s = r\alpha$, linear velocity is

    $v = 6.25(90\pi) \approx 918.9$ in./min.

35. The radius of the circle is $r = 424/(2\pi)$. Applying the formula $s = r\alpha$, the linear velocity is

$$v = \left(\frac{424 \text{ m}}{2\pi}\right) \cdot \frac{2\pi}{30 \text{ min.}} \cdot \frac{1 \text{ min.}}{60 \text{ sec}} \cdot \frac{3.28 \text{ ft}}{1 \text{ m}} \approx$$

0.8 ft/sec.

37. The radius of the blade is

$$10 \text{ in.} = 10 \cdot \frac{1}{12 \cdot 5280} \approx 0.0001578 \text{ miles.}$$

Since the angle rotated in one hour is

$$\alpha = 2800 \cdot 2\pi \cdot 60 = 336,000\pi,$$

the linear velocity is

$$v = r\alpha \approx (0.0001578) \cdot 336,000\pi \approx 166.6 \text{ mph.}$$

39. In 1 hr, the saw rotates through an angle $\alpha = 3450(60) \cdot 2\pi$. Note $s = r\alpha$ is a formula for the arc length. Then the difference in the linear velocity is

$$\frac{6}{12}\alpha - \frac{5}{12}\alpha \text{ or } \frac{3450(60) \cdot 2\pi}{(3600)12} \approx 30.1 \text{ ft/sec.}$$

41. The angular velocity of any point on the surface of the earth is $w = \dfrac{\pi}{12}$ rad/hr.

A point 1 mile from the North Pole is approximately 1 mile from the axis of the earth. The linear velocity of that point is $v = w \cdot r = \dfrac{\pi}{12} \cdot 1 \approx 0.3$ mph.

43. Since $\alpha = 15° \approx 0.261799$ radians, the linear velocity is $v = \dfrac{r\alpha}{t} = \dfrac{r\alpha}{1} = r\alpha \approx$
6.5(3950) · 0.261799 ≈ 6721.7 mph.

45. The linear velocity is given by $v = r\omega$.

If $v = 2$ ft/sec, then the angular velocity is $\omega = \frac{2}{r}$ radians/sec.

As the radius increases, the angular velocity decreases.

47. Since the velocity at point A is 10 ft/sec, the linear velocity at B and C are both 10 ft/sec. The angular velocity at C is

$$\omega = \frac{v}{r} = \frac{10 \text{ ft/sec}}{5/12 \text{ ft}} = 24 \text{ rad/sec}$$

and the angular velocity at B is

$$\omega = \frac{v}{r} = \frac{10 \text{ ft/sec}}{3/12 \text{ ft}} = 40 \text{ rad/sec.}$$

49. Since the chain ring which has 52 teeth turns at the rate of 1 rev/sec, the cog with 26 teeth will turn at the rate of 2 rev/sec. Thus, the linear velocity of the bicycle with 13.5-in.-radius wheels is

$$\frac{2\text{rev}}{\text{sec}} \cdot \frac{2\pi(13.5)\text{in.}}{\text{rev}} \cdot \frac{1\text{mile}}{63,360\text{in.}} \cdot \frac{3600\text{sec}}{1\text{hr}} \approx 9.6 \text{ mph.}$$

53. a) $12°65'96'' = 12°66'36'' = 13°6'36''$

b) $27°77'68'' - 9°19'29'' = 18°58'39''$

55. 210°

57. $A = \pi r^2$

59. a) Let t be a fraction of an hour, i.e., $0 \le t \le 1$. If the angle between the hour and minute hands is 120°, then

$$\begin{aligned} 360t - 30t &= 120 \\ 330t &= 120 \\ t &= \frac{12}{33}\text{hr} \\ t &\approx 21 \text{ min}, 49.1 \text{ sec.} \end{aligned}$$

Thus, the hour and minute hands will be 120° apart when the time is 12:21:49.1. Moreover, this is the only time between 12 noon and 1 pm that the angle is 120°.

b) We measure angles clockwise from the 12 0'clock position.
At 12:21:49.1, the hour hand is pointing to the 10.9°-angle, the minute hand is pointing to the 130.9°-angle, and the second hand is pointing to the 294.5°-angle. Thus, the three hands of the clock cannot divide the face of the clock into thirds.

c) No, as discussed in part b).

For Thought

1. True, since $(0, 1)$ is on the positive y-axis and
$$\sin 90° = \frac{y}{r} = \frac{1}{1} = 1.$$

2. True, since $(1, 0)$ is on the positive x-axis and
$$\sin 0° = \frac{x}{r} = \frac{0}{1} = 0.$$

3. True, since $(1, 1)$ is on the terminal side of $45°$ and $r = \sqrt{2}$, and so
$$\cos 45° = \frac{x}{r} = \frac{1}{\sqrt{2}}.$$

4. True, since $(1, 1)$ is on the terminal side of $45°$ we get
$$\tan 45° = \frac{y}{x} = \frac{1}{1} = 1.$$

5. True, since $(\sqrt{3}, 1)$ is on the terminal side of $30°$ and $r = 2$ we find
$$\sin 30° = \frac{y}{r} = \frac{1}{2}.$$

6. True, since $(0, 1)$ is on the positive y-axis we get
$$\cos(\pi/2) = \frac{x}{r} = \frac{0}{1} = 1.$$

7. True, since the terminal sides of $390°$ and $30°$ are the same.

8. False, since $\sin(-\pi/3) = -\frac{\sqrt{3}}{2}$ and
$$\sin(\pi/3) = \frac{\sqrt{3}}{2}.$$

9. True, since $\csc \alpha = \frac{1}{\sin \alpha} = \frac{1}{1/5} = 5.$

10. True, since $\sec \alpha = \frac{1}{\cos \alpha} = \frac{1}{2/3} = 1.5.$

1.4 Exercises

1. $\sin \alpha$, $\cos \alpha$

3. reciprocals

5. Note $r = \sqrt{1^2 + 2^2} = \sqrt{5}$. Then
$$\sin \alpha = \frac{y}{r} = \frac{2}{\sqrt{5}} = \frac{2\sqrt{5}}{5},$$
$$\cos \alpha = \frac{x}{r} = \frac{1}{\sqrt{5}} = \frac{\sqrt{5}}{5},$$
$$\tan \alpha = \frac{y}{x} = \frac{2}{1} = 2, \ \csc \alpha = \frac{1}{\sin \alpha} = \frac{\sqrt{5}}{2},$$
$$\sec \alpha = \frac{1}{\cos \alpha} = \frac{\sqrt{5}}{1} = \sqrt{5}, \text{ and}$$
$$\cot \alpha = \frac{1}{\tan \alpha} = \frac{1}{2}.$$

7. Note $r = \sqrt{0^2 + 1^2} = 1$. Then
$$\sin \alpha = \frac{y}{r} = \frac{1}{1} = 1, \ \cos \alpha = \frac{x}{r} = \frac{0}{1} = 0,$$
$$\tan \alpha = \frac{y}{x} = \frac{1}{0} = \text{undefined},$$
$$\csc \alpha = \frac{1}{\sin \alpha} = \frac{1}{1} = 1,$$
$$\sec \alpha = \frac{1}{\cos \alpha} = -\frac{1}{0} = \text{undefined, and}$$
$$\cot \alpha = \frac{x}{y} = \frac{0}{1} = 0.$$

9. Note $r = \sqrt{1^2 + 1^2} = \sqrt{2}$. Then
$$\sin \alpha = \frac{y}{r} = \frac{1}{\sqrt{2}} = \frac{\sqrt{2}}{2},$$
$$\cos \alpha = \frac{x}{r} = \frac{1}{\sqrt{2}} = \frac{\sqrt{2}}{2},$$
$$\tan \alpha = \frac{y}{x} = \frac{1}{1} = 1,$$
$$\csc \alpha = \frac{1}{\sin \alpha} = \frac{1}{1/\sqrt{2}} = \sqrt{2},$$
$$\sec \alpha = \frac{1}{\cos \alpha} = \frac{1}{1/\sqrt{2}} = \sqrt{2}, \text{ and}$$
$$\cot \alpha = \frac{1}{\tan \alpha} = \frac{1}{1} = 1.$$

11. Note $r = \sqrt{(-2)^2 + 2^2} = 2\sqrt{2}$. Then
$$\sin \alpha = \frac{y}{r} = \frac{2}{2\sqrt{2}} = \frac{\sqrt{2}}{2},$$
$$\cos \alpha = \frac{x}{r} = \frac{-2}{2\sqrt{2}} = -\frac{\sqrt{2}}{2},$$
$$\tan \alpha = \frac{y}{x} = \frac{2}{-2} = -1,$$

$$\csc\alpha = \frac{1}{\sin\alpha} = \frac{1}{2/(2\sqrt{2})} = \sqrt{2},$$

$$\sec\alpha = \frac{1}{\cos\alpha} = \frac{1}{-2/(2\sqrt{2})} = -\sqrt{2}, \text{ and}$$

$$\cot\alpha = \frac{1}{\tan\alpha} = \frac{1}{-1} = -1.$$

13. Note $r = \sqrt{(-4)^2 + (-6)^2} = 2\sqrt{13}$. Then

$$\sin\alpha = \frac{y}{r} = \frac{-6}{2\sqrt{13}} = -\frac{3\sqrt{13}}{13},$$

$$\cos\alpha = \frac{x}{r} = \frac{-4}{2\sqrt{13}} = \frac{-2\sqrt{13}}{13},$$

$$\tan\alpha = \frac{y}{x} = \frac{-6}{-4} = \frac{3}{2},$$

$$\csc\alpha = \frac{1}{\sin\alpha} = -\frac{13}{3\sqrt{13}} = -\frac{\sqrt{13}}{3},$$

$$\sec\alpha = \frac{1}{\cos\alpha} = -\frac{13}{2\sqrt{13}} = -\frac{\sqrt{13}}{2}, \text{ and}$$

$$\cot\alpha = \frac{1}{\tan\alpha} = \frac{2}{3}.$$

15. 0 **17.** -1

19. 0

21. Undefined

23. -1 **25.** -1

27. $\dfrac{\sqrt{2}}{2}$ **29.** $\dfrac{\sqrt{2}}{2}$

31. -1 **33.** $\sqrt{2}$

35. $\dfrac{1}{2}$ **37.** $\dfrac{1}{2}$

39. $-\dfrac{\sqrt{3}}{3}$ **41.** -2

43. 2 **45.** $\sqrt{3}$

47. $\dfrac{\cos(\pi/3)}{\sin(\pi/3)} = \dfrac{1/2}{\sqrt{3}/2} = \dfrac{\sqrt{3}}{3}$

49. $\dfrac{\sin(7\pi/4)}{\cos(7\pi/4)} = \dfrac{-\sqrt{2}/2}{\sqrt{2}/2} = -1$

51. $\sin\left(\dfrac{\pi}{3} + \dfrac{\pi}{6}\right) = \sin\left(\dfrac{\pi}{2}\right) = 1$

53. $\dfrac{1 - \cos(5\pi/6)}{\sin(5\pi/6)} = \dfrac{1 - (-\sqrt{3}/2)}{1/2} \cdot \dfrac{2}{2} = 2 + \sqrt{3}$

55. $\dfrac{\sqrt{2}}{2} + \dfrac{\sqrt{2}}{2} = \sqrt{2}$

57. $\cos(45°)\cos(60°) - \sin(45°)\sin(60°) =$

$$\frac{\sqrt{2}}{2} \cdot \frac{1}{2} - \frac{\sqrt{2}}{2} \cdot \frac{\sqrt{3}}{2} = \frac{\sqrt{2} - \sqrt{6}}{4}$$

59. $\dfrac{1 - \cos(\pi/3)}{2} = \dfrac{1 - 1/2}{2} = \dfrac{1}{4}.$

61. $2\cos(210°) = 2 \cdot \dfrac{-\sqrt{3}}{2} = -\sqrt{3}.$

63. 0.6820

65. -0.6366

67. 0.0105

69. $\dfrac{1}{\sin(23°48')} \approx 2.4780$

71. $\dfrac{1}{\cos(-48°3'12'')} \approx 1.4960$

73. $\dfrac{1}{\tan(\pi/9)} \approx 2.7475$

75. 0.8578

77. 0.2679

79. 0.9894

81. 2.9992

83. $\sin(2 \cdot \pi/4) = \sin(\pi/2) = 1$

85. $\cos(2 \cdot \pi/6) = \cos(\pi/3) = \dfrac{1}{2}$

87. $\sin((3\pi/2)/2) = \sin(3\pi/4) = \dfrac{\sqrt{2}}{2}$

89.

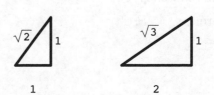

91. $\csc \alpha = \dfrac{1}{\sin \alpha} = \dfrac{1}{3/4} = \dfrac{4}{3}$

93. $\cos \alpha = \dfrac{1}{\sec \alpha} = \dfrac{1}{10/3} = \dfrac{3}{10}$

95. a) II, since $y > 0$ and $x < 0$ in Quadrant II

    b) IV, since $y < 0$ and $x > 0$ in Quadrant IV

    c) III, since $y/x > 0$ and $x < 0$ in Quadrant III

    d) II, since $y/x < 0$ and $y > 0$ in Quadrant II

97. If $R = 3$, $r = 14$, and $\theta = 60°$, then

$$
\begin{aligned}
T &= \frac{14 \cos 60° - 3}{\sin 60°} \\
&= \frac{14(1/2) - 3}{\sqrt{3}/2} \\
&\approx 4.6 \text{ inches}
\end{aligned}
$$

99. $180° - 2\,(9°38'52'') = 180° - (19°17'44'') =$
$179°59'60'' - 19°17'44'' = 160°42'16''$

101. $\dfrac{13\pi}{12} \cdot \dfrac{180°}{\pi} = 195°$

103. The radius is 15 inches. The linear velocity is

$$v = \omega r = 2000(2\pi) \cdot 15 \text{ inches/minute}$$

Equivalently,

$$v = 2000(2\pi) \cdot 15 \frac{60}{12(5280)} \approx 178.5 \text{ mph}$$

105. Let P stand for Porsche, N for Nissan, and C for Chrysler. The fifteen preferences could be the following:

Six preferences: P, N, C (1st, 2nd, 3rd respectively)

Five preferences: C, N, P

Three preferences: N, C, P

One preference: N, P, C

For Thought

1. True

2. False, rather $\alpha = 30°$ since $\cos 30° = \dfrac{\sqrt{3}}{2}$.

3. False, since $\sin^{-1}(\sqrt{2}/2) = 45°$.

4. False, since $\cos^{-1}(1/2) = 60°$.

5. False, since $\tan^{-1}(1) = 45°$.

6. False, since $c = \sqrt{2^2 + 4^2} = \sqrt{20}$.

7. True, since $c = \sqrt{3^2 + 4^2} = \sqrt{25} = 5$.

8. False, since $c = \sqrt{4^2 + 8^2} = \sqrt{80}$.

9. True, since $\alpha + \beta = 90°$.

10. False, otherwise $1 = \sin 90° = \dfrac{\text{hyp}}{\text{adj}}$ and we find hyp $=$ adj, which is impossible. The hypotenuse is longer than each of the legs of a right triangle.

1.5 Exercises

1. right

3. inverse sine

5. adjacent side, hypotenuse

7. $45°$ **9.** $60°$

11. $60°$ **13.** $0°$

15. $83.6°$

17. $67.6°$

19. $26.1°$

21. $29.1°$

23. Note, the hypotenuse is hyp $= \sqrt{13}$.

Then $\sin \alpha = \dfrac{\text{opp}}{\text{hyp}} = \dfrac{2}{\sqrt{13}} = \dfrac{2\sqrt{13}}{13}$,

$\cos \alpha = \dfrac{\text{adj}}{\text{hyp}} = \dfrac{3}{\sqrt{13}} = \dfrac{3\sqrt{13}}{13}$,

$\tan \alpha = \dfrac{\text{opp}}{\text{adj}} = \dfrac{2}{3}$,

$$\csc \alpha = \frac{\text{hyp}}{\text{opp}} = \frac{\sqrt{13}}{2},$$

$$\sec \alpha = \frac{\text{hyp}}{\text{adj}} = \frac{\sqrt{13}}{3}, \text{ and}$$

$$\cot \alpha = \frac{\text{adj}}{\text{opp}} = \frac{3}{2}.$$

25. Note, the hypotenuse is $4\sqrt{5}$.
Then $\sin(\alpha) = \sqrt{5}/5, \cos(\alpha) = 2\sqrt{5}/5,$
$\tan(\alpha) = 1/2, \sin(\beta) = 2\sqrt{5}/5,$
$\cos(\beta) = \sqrt{5}/5,$ and $\tan(\beta) = 2.$

27. Note, the hypotenuse is $2\sqrt{34}$.
$\sin(\alpha) = 3\sqrt{34}/34, \cos(\alpha) = 5\sqrt{34}/34,$
$\tan(\alpha) = 3/5, \sin(\beta) = 5\sqrt{34}/34,$
$\cos(\beta) = 3\sqrt{34}/34,$ and $\tan(\beta) = 5/3.$

29. Note, the side adjacent to β has length 12.
Then $\sin(\alpha) = 4/5, \cos(\alpha) = 3/5,$
$\tan(\alpha) = 4/3, \sin(\beta) = 3/5,$
$\cos(\beta) = 4/5,$ and $\tan(\beta) = 3/4.$

31. Form the right triangle with $a = 6, b = 10$.

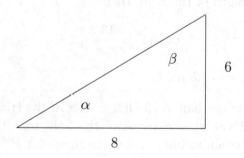

Note: $c = \sqrt{6^2 + 8^2} = 10, \tan(\alpha) = 6/8,$
so $\alpha = \tan^{-1}(6/8) \approx 36.9°$ and $\beta \approx 53.1°$.

33. Form the right triangle with $b = 6, c = 8.3$.

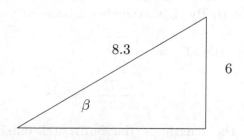

Note: $a = \sqrt{8.3^2 - 6^2} \approx 5.7, \sin(\beta) = 6/8.3,$
so $\beta = \sin^{-1}(6/8.3) \approx 46.3°$ and $\alpha \approx 43.7°$.

35. Form the right triangle with $\alpha = 16°, c = 20$.

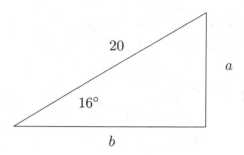

Since $\sin(16°) = a/20$ and $\cos(16°) = b/20,$
$a = 20 \sin(16°) \approx 5.5$ and
$b = 20 \cos(16°) \approx 19.2$. Also $\beta = 74°.$

37. Form the right triangle with $\alpha = 39°9', a = 9$.

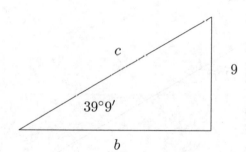

Since $\sin(39°9') = 9/c$ and $\tan(39°9') = 9/b,$
then $c = 9/\sin(39°9') \approx 14.3$ and
$b = 9/\tan(39°9') \approx 11.1$. Also $\beta = 50°51'.$

39. Let h be the height of the buliding.

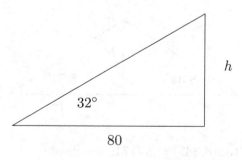

Since $\tan(32°) = h/80$, we obtain

$$h = 80 \cdot \tan(32°) \approx 50 \text{ ft.}$$

41. Let x be the distance between Muriel and the road at the time she encountered the swamp.

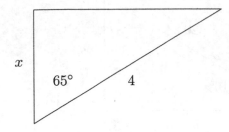

Since $\cos(65°) = x/4$, we obtain $x = 4 \cdot \cos(65°) \approx 1.7$ miles.

43. Let x be the distance between the car and a point on the highway directly below the overpass.

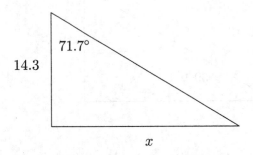

Since $\tan(71.7°) = x/14.3$, we obtain $x = 14.3 \cdot \tan(71.7°) \approx 43.2$ meters.

45. Let h be the height as in the picture below.

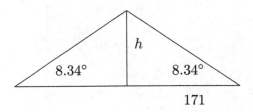

Since $\tan(8.34°) = h/171$, we obtain $h = 171 \cdot \tan(8.34°) \approx 25.1$ ft.

47. Let α be the angle the guy wire makes with the ground.

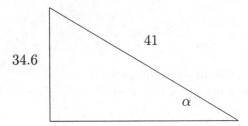

From the Pythagorean Theorem, the distance of the point to the base of the antenna is

$$\sqrt{41^2 - 34.6^2} \approx 22 \text{ meters.}$$

Also, $\alpha = \sin^{-1}(34.6/41) \approx 57.6°$.

49. Consider a right triangle with a height of 13.3 in., and the base is the length of the trail. Let $64°$ be the angle opposite the height. If x is the length of the trail, then

$$\tan 64° = \frac{13.3}{x}.$$

Thus, $x \approx 6.5$ in.

51. Choose the ball in the left corner of the triangle. From the center A of the ball draw the perpendicular line segment to the point B below and on the outside perimeter of the rack.

Since the diameter of the ball is 2.25 in., and the sides of the triangular rack is 0.25 in., we obtain $AB = 2.25/2 + 0.25 = 1.375$ in.

Let x be the length from the left vertex of the rack to B. By right triangle trigonometry,

$$\tan 30° = \frac{1.375}{x}$$

$$x = \frac{1.375}{\tan 30°} \approx 2.38.$$

Then the length of the horizontal outside perimeter of the rack is $2.25 \times 4 + 2x$. The total length of the outside perimeter is

$$3\left(2.25(4) + 2x\right) \approx 41.3 \text{ in.}$$

53. Draw a line segment from the center A of the pentagon to the midpoint C of one of the sides of the regular pentagon. This line segment should be perpendicular to the side. Consider a right triangle with vertices at A, C, and an adjacent vertex B of the hexagon near C. The angle at A of the right triangle is $36°$, and the hypotenuse r is the distance from the center A to vertex B. Using right triangle trigonometry, we obtain

$$r = \frac{1}{\sin 36°} \approx 1.701.$$

Furthermore, the side adjacent to $36°$ is $h - r$. Applying right triangle trigonometry, we have

$$\cos 36° = \frac{h - r}{r}$$

Solving for h, we find

$$h = r(1 + \cos 36°) \approx 3.08 \text{ m}$$

55. Note, 1.75 sec. $= 1.75/3600$ hour.

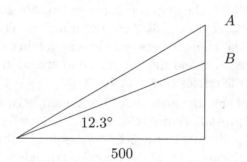

The distance in miles between A and B is

$$\frac{500 \left(\tan(15.4°) - \tan(12.3°) \right)}{5280} \approx 0.0054366$$

The speed is

$$\frac{0.0054366}{(1.75/3600)} \approx 11.2 \text{ mph}$$

and the car is not speeding.

57. Let h be the height.

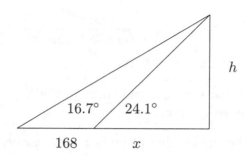

Note, $\tan 24.1° = \dfrac{h}{x}$ and $\tan 16.7° = \dfrac{h}{168 + x}$.

Solve for h in the second equation and substitute $x = \dfrac{h}{\tan 24.1°}$.

$$h = \tan(16.7°) \cdot \left(168 + \frac{h}{\tan 24.1°} \right)$$

$$h \quad - \frac{h \tan(16.7°)}{\tan 24.1°} = \tan(16.7°) \cdot 168$$

$$h \quad = \quad \frac{168 \cdot \tan(16.7°)}{1 - \tan(16.7°)/\tan(24.1°)}$$

$$h \quad \approx \quad 153.1 \text{ meters}$$

The height is 153.1 meters.

59. Let y be the closest distance the boat can come to the lighthouse LH.

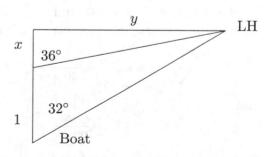

Since $\tan(36°) = y/x$ and $\tan(32°) = y/(1 + x)$, we obtain

$$\tan(32°) = \frac{y}{1 + y/\tan(36°)}$$

$$\tan(32°) \quad + \frac{\tan(32°)y}{\tan(36°)} = y$$

$$\tan(32°) = y\left(1 - \frac{\tan(32°)}{\tan(36°)}\right)$$

$$y = \frac{\tan(32°)}{1 - \tan(32°)/\tan(36°)}$$

$$y \approx 4.5 \text{ km}.$$

The closest the boat will come to the lighthouse is 4.5 km.

61. Let x be the number of miles in one parsec.

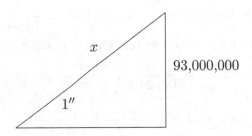

Since $\sin(1'') = \dfrac{93,000,000}{x}$, we obtain

$$x = \frac{93,000,000}{\sin(1/3600°)} \approx 1.9 \times 10^{13} \text{ miles}.$$

Light travels one parsec in 3.26 years since

$$\frac{x}{193,000(63,240)} \approx 3.26 \text{ years}.$$

63. Let h be the height of the building, and let x be the distance between C and the building. Using right triangle trigonometry, we obtain

$$\frac{1}{\sqrt{3}} = \tan 30° = \frac{h}{40+x}$$

and

$$1 = \tan 45° = \frac{h}{20+x}.$$

Solving simultaneously, we find $x = 10(\sqrt{3}-1)$ and $h = 10(\sqrt{3}+1)$. However,

$$\tan C = \frac{h}{x} = \frac{10(\sqrt{3}+1)}{10(\sqrt{3}-1)} = 2+\sqrt{3}$$

Since $\tan 75° = 2+\sqrt{3}$ by the addition formula for tangent, we obtain

$$C = 75°$$

65. a) The distance from a center to the nearest vertex of the square is $6\sqrt{2}$ by the Pythagorean theorem. Then the diagonal of the square is $12 + 12\sqrt{2}$. From which, the side of the square is $12 + 6\sqrt{2}$ by the Pythagorean theorem as shown below.

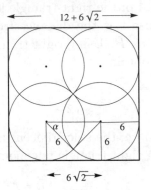

The area A_c of the region in one corner of the box that is *not* watered is obtained by subtracting one-fourth of the area of a circle of radius 6 from the area of a 6-by-6 square:

$$A_c = 36 - 9\pi.$$

Note, the distance between two horizontal centers is $6\sqrt{2}$ as shown above. Then the angle α between the line joining the centers and the line to the intersection of the circles is $\alpha = \pi/4$.

Thus, the area A_b of the region between two adjacent circles that is not watered is the area of a 6-by-$6\sqrt{2}$ square minus the area, 18, of the isosceles triangle with base angle $\alpha = \pi/4$, and minus the combined area 9π of two sectors with central angle $\pi/4$:

$$A_b = 36\sqrt{2} - 18 - 9\pi$$

Hence, the total area not watered is

$$4(A_c + A_b) = 4\left(36 - 9\pi + 36\sqrt{2} - 18 - 9\pi\right)$$
$$= 72\left(1 + 2\sqrt{2} - \pi\right) \text{ m}^2$$

b) Since the side of the square is $12 + 6\sqrt{2}$, the area that is watered by at least one sprinkler is

$$100\% - \frac{4(A_c + A_b)}{(12+6\sqrt{2})^2} \cdot 100\% \approx 88.2\%$$

67. In the triangle below CE stands for the center of the earth, and LS is a point on the surface of the earth lying in the line of sight of the satellite.

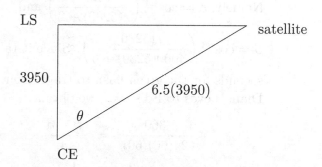

Since $\cos(\theta) = \dfrac{3950}{6.5(3950)} = 1/6.5$, we get

$\theta = \cos^{-1}(1/6.5) \approx 1.41634$ radians. But 2θ is the widest angle formed by a sender and receiver of a signal with vertex CE. The maximum distance is the arclength subtended by 2θ, i.e.,

$s = r \cdot 2\theta = 3950 \cdot 2 \cdot 1.41634 \approx 11,189$ miles.

69. First, consider the figure below.

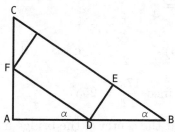

Suppose $AC = s$, $AB = \dfrac{3s}{2}$, $BD = h$, $DE = w$, and $DF = L$. Note, $\tan \alpha = \dfrac{AC}{AB} = \dfrac{2}{3}$. Since $\sin \alpha = \dfrac{2}{\sqrt{13}}$ and $\sin \alpha = \dfrac{DE}{BD}$, we obtain $w = \dfrac{2h}{\sqrt{13}}$. Since $\cos \alpha = \dfrac{AD}{DF}$, we get

$$\cos \alpha = \dfrac{\dfrac{3s}{2} - h}{L}.$$

Solving for L, we find $L = \dfrac{\dfrac{3s}{2} - h}{\cos \alpha}$ and since

$\cos \alpha = \dfrac{3}{\sqrt{13}}$, we get

$$L = \frac{\sqrt{13}}{3}\left(\frac{3s}{2} - \frac{w\sqrt{13}}{2}\right).$$

So, the area of the parking lot is

$$wL = \frac{w\sqrt{13}}{3}\left(\frac{3s}{2} - \frac{w\sqrt{13}}{2}\right).$$

Since this area represents a quadratic function of w, one can find the vertex of its graph and conclude that the maximum area of the rectangle is obtained if one chooses $w = \dfrac{3s}{2\sqrt{13}}$.

Correspondingly, we obtain $L = \dfrac{s\sqrt{13}}{4}$.

Finally, given $s = 100$ feet, the dimensions of the house with maximum area are

$$w = \frac{3(100)}{2\sqrt{13}} \approx 41.60 \text{ ft and } L = \frac{100\sqrt{13}}{4} \approx 90.14 \text{ ft.}$$

71. Consider the right triangle formed by the hook, the center of the circle, and a point on the circle where the chain is tangent to the circle.

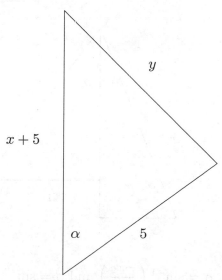

Then $\tan \alpha = \dfrac{y}{5}$ or $y = 5 \tan \alpha$. Since the chain is 40 ft long and the angle $2\pi - 2\alpha$ intercepts an arc around the pipe where the chain wraps around the circle, we obtain

$$2y + 5(2\pi - 2\alpha) = 40.$$

By substitution, we get

$$10\tan\alpha + 10\pi - 10\alpha = 40.$$

With a graphing calculator, we obtain $\alpha \approx 1.09835$ radians. From the figure above, we get $\cos\alpha = \dfrac{5}{5+x}$. Solving for x, we obtain

$$x = \frac{5 - 5\cos\alpha}{\cos\alpha} \approx 5.987 \text{ ft.}$$

73. Assume the circle is given by

$$x^2 + (y-r)^2 = r^2$$

where r is the radius. Suppose the points where the blocks touch the circle are at the points $(-x_2, 1)$ and $(x_1, 2)$ where $x_1, x_2 > 0$. Substitute the points into the equation of the circle. Then

$$x_1^2 + (2-r)^2 = r^2$$

and

$$x_2^2 + (1-r)^2 = r^2.$$

From which we obtain $x_1^2 + 4 - 4r = 0$ and $x_2^2 + 1 - 2r = 0$. Thus, $x_1 = \sqrt{4r-4}$ and $x_2 = -\sqrt{2r-1}$.

Consider the two triangles below.

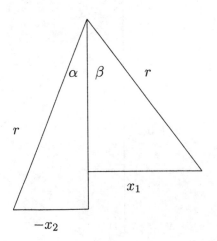

Note, $\alpha = \sin^{-1}\left(\dfrac{-x_2}{r}\right)$ and $\beta = \sin^{-1}\left(\dfrac{x_1}{r}\right)$.

Since 6 ft is the arclength between the blocks, we have $r(\alpha + \beta) = 6$. By substitution, we obtain

$$r\left(\sin^{-1}\left(\frac{\sqrt{2r-1}}{r}\right) + \sin^{-1}\left(\frac{\sqrt{4r-4}}{r}\right)\right) = 6$$

With the aid of a graphing calculator, we find $r \approx 2.768$ ft.

75. Let r be the radius of the earth. Consider the angles shown in Exercises 68 a) and b).

Namely, $\alpha = \cos^{-1}\left(\dfrac{r(5280)}{r(5280) + 2}\right)$ and

$\beta = \cos^{-1}\left(\dfrac{r(5280)}{4000(5280) + 6}\right)$. Since it takes 4 seconds for the green flash to travel from Diane's eyes to Ed's eyes, we obtain

$$\frac{360}{24(60)(60)} = \frac{\beta - \alpha}{4}$$

$$\beta - \alpha = \frac{1}{60}.$$

With a calculator, we get that the solution to

$$\cos^{-1}\left(\frac{r(5280)}{4000(5280) + 6}\right) - \cos^{-1}\left(\frac{r(5280)}{r(5280) + 2}\right) = \frac{1}{60}$$

is $r \approx 4798$ miles.

79. Let $r = \sqrt{4^2 + 3^2} = 5$

a) $\sin\alpha = \dfrac{y}{r} = \dfrac{3}{5}$

b) $\cos\alpha = \dfrac{x}{r} = \dfrac{4}{5}$

c) $\tan\alpha = \dfrac{y}{x} = \dfrac{3}{4}$

81. a) $\dfrac{\sqrt{3}}{2}$ b) $-\dfrac{\sqrt{3}}{2}$ c) $-\sqrt{3}$

83. Quadrant III since $\dfrac{17\pi}{12} = 255°$

85. We assume that the center of the bridge moves straight upward to the center of the arc where the bridge expands. The arc has length $100 + \frac{1}{12}$ feet. To find the central angle α, we use $s = r\alpha$ and $\sin(\alpha/2) = 50/r$. Then

$$\alpha = 2\sin^{-1}(50/r).$$

Solving the equation

$$100 + \frac{1}{12} = 2r\sin^{-1}\left(\frac{50}{r}\right)$$

with a graphing calculator, we find

$$r \approx 707.90241 \text{ ft.}$$

The distance from the chord to the center of the circle can be found by using the Pythagorean theorem, and it is

$$d = \sqrt{r^2 - 50^2} \approx 706.1344221 \text{ ft}.$$

Then the distance from the center of the arc to the cord is

$$r - d \approx 1.767989 \text{ ft} \approx 21.216 \text{ inches}.$$

For Thought

1. True, since $\sin^2 \alpha + \cos^2 \alpha = 1$ for any real number α.

2. True

3. False, since α is in Quadrant IV.

4. True

5. False, rather $\sin \alpha = -\dfrac{1}{2}$.

6. True   7. True

8. False, since the reference angle is $\dfrac{\pi}{3}$.

9. True, since $\cos 120° = -\dfrac{1}{2} = -\cos 60°$.

10. True, since $\sin(7\pi/6) = -\dfrac{1}{2} = -\sin(\pi/6)$.

1.6 Exercises

1. fundamental

3. $\sin \alpha = \pm\sqrt{1 - \cos^2 \alpha} = \pm\sqrt{1 - 1^2} = 0$

5. Use the Fundamental Identity.

$$\left(\frac{5}{13}\right)^2 + \cos^2(\alpha) = 1$$
$$\frac{25}{169} + \cos^2(\alpha) = 1$$
$$\cos^2(\alpha) = \frac{144}{169}$$
$$\cos(\alpha) = \pm\frac{12}{13}$$

Since α is in quadrant II, $\cos(\alpha) = -12/13$.

7. Use the Fundamental Identity.

$$\left(\frac{3}{5}\right)^2 + \sin^2(\alpha) = 1$$
$$\frac{9}{25} + \sin^2(\alpha) = 1$$
$$\sin^2(\alpha) = \frac{16}{25}$$
$$\sin(\alpha) = \pm\frac{4}{5}$$

Since α is in quadrant IV, $\sin(\alpha) = -4/5$.

9. Use the Fundamental Identity.

$$\left(\frac{1}{3}\right)^2 + \cos^2(\alpha) = 1$$
$$\frac{1}{9} + \cos^2(\alpha) = 1$$
$$\cos^2(\alpha) = \frac{8}{9}$$
$$\cos(\alpha) = \pm\frac{2\sqrt{2}}{3}$$

Since $\cos(\alpha) > 0$, $\cos(\alpha) = \dfrac{2\sqrt{2}}{3}$.

11. $30°, \pi/6$

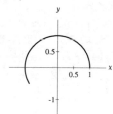

13. $60°, \pi/3$

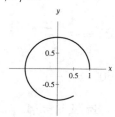

15. $60°, \pi/3$

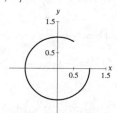

17. $30°, \pi/6$

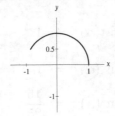

19. $45°, \pi/4$

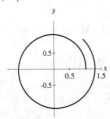

21. $45°, \pi/4$

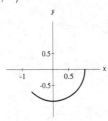

23. $\sin(135°) = \sin(45°) = \dfrac{\sqrt{2}}{2}$

25. $\cos\left(\dfrac{5\pi}{3}\right) = \cos\left(\dfrac{\pi}{3}\right) = \dfrac{1}{2}$

27. $\sin\left(\dfrac{7\pi}{4}\right) = -\sin\left(\dfrac{\pi}{4}\right) = -\dfrac{\sqrt{2}}{2}$

29. $\cos\left(-\dfrac{17\pi}{6}\right) = -\cos\left(\dfrac{\pi}{6}\right) = -\dfrac{\sqrt{3}}{2}$

31. $\sin\left(-45°\right) = -\sin\left(45°\right) = -\dfrac{\sqrt{2}}{2}$

33. $\cos\left(-240°\right) = -\cos\left(60°\right) = -\dfrac{1}{2}$

35. The reference angle of $3\pi/4$ is $\pi/4$.

Then $\sin(3\pi/4) = \sin(\pi/4) = \dfrac{\sqrt{2}}{2}$,

$\cos(3\pi/4) = -\cos(\pi/4) = -\dfrac{\sqrt{2}}{2}$,

$\tan(3\pi/4) = -\tan(\pi/4) = -1$,

$\csc(3\pi/4) = \csc(\pi/4) = \sqrt{2}$,

$\sec(3\pi/4) = -\sec(\pi/4) = -\sqrt{2}$, and

$\cot(3\pi/4) = -\cot(\pi/4) = -1.$

37. The reference angle of $4\pi/3$ is $\pi/3$.

Then $\sin(4\pi/3) = -\sin(\pi/3) = -\dfrac{\sqrt{3}}{2}$,

$\cos(4\pi/3) = -\cos(\pi/3) = -\dfrac{1}{2}$,

$\tan(4\pi/3) = \tan(\pi/3) = \sqrt{3}$,

$\csc(4\pi/3) = -\csc(\pi/3) = -\dfrac{2\sqrt{3}}{3}$,

$\sec(4\pi/3) = -\sec(\pi/3) = -2$, and

$\cot(4\pi/3) = \cot(\pi/3) = \dfrac{\sqrt{3}}{3}.$

39. The reference angle of $300°$ is $60°$.

Then $\sin(300°) = -\sin(60°) = -\dfrac{\sqrt{3}}{2}$,

$\cos(300°) = \cos(60°) = \dfrac{1}{2}$,

$\tan(300°) = -\tan(60°) = -\sqrt{3}$,

$\csc(300°) = -\csc(60°) = -\dfrac{2\sqrt{3}}{3}$,

$\sec(300°) = \sec(60°) = 2$, and

$\cot(300°) = -\cot(60°) = -\dfrac{\sqrt{3}}{3}.$

41. The reference angle of $-135°$ is $45°$.

Then $\sin(-135°) = -\sin(45°) = -\dfrac{\sqrt{2}}{2}$,

$\cos(-135°) = -\cos(45°) = -\dfrac{\sqrt{2}}{2}$,

$\tan(-135°) = \tan(45°) = 1$,

$\csc(-135°) = -\csc(45°) = -\sqrt{2}$,

$\sec(-135°) = -\sec(45°) = -\sqrt{2}$, and

$\cot(-135°) = \cot(45°) = 1.$

43. False, since $\sin 210° = -\sin 30°$.

45. True, since $\cos 330° = \dfrac{\sqrt{3}}{2} = \cos 30°$.

47. False, since $\sin 179° = \sin 1°$.

49. True, for the reference angle is $\pi/7$, $6\pi/7$ is in Quadrant II, and cosine is negative in Quadrant II.

51. False, since $\sin(23\pi/24) = \sin(\pi/24)$.

53. True, for the reference angle is $\pi/7$, $13\pi/7$ is in Quadrant IV, and cosine is positive in Quadrant IV.

55. If $h = 18$, then

$$
\begin{aligned}
T &= 18\sin\left(\frac{\pi}{12}(6)\right) + 102 \\
&= 18\sin\left(\frac{\pi}{2}\right) + 102 = 18 + 102 \\
&= 120°\text{F}.
\end{aligned}
$$

If $h = 6$, then

$$
\begin{aligned}
T &= 18\sin\left(\frac{\pi}{12}(-6)\right) + 102 \\
&= 18\sin\left(-\frac{\pi}{2}\right) + 102 = -18 + 102 \\
&= 84°\text{F}.
\end{aligned}
$$

57. Note, $x(t) = 4\sin(t) + 3\cos(t)$.

   a) Initial position is

   $$x(0) = 3\cos 0 = 3.$$

   b) If $t = 5\pi/4$, the position is

   $$
   \begin{aligned}
   x(5\pi/4) &= 4\sin(5\pi/4) + 3\cos(5\pi/4) \\
   &= -2\sqrt{2} - \frac{3\sqrt{2}}{2} \\
   &= -\frac{7\sqrt{2}}{2}.
   \end{aligned}
   $$

59. The angle between the tips of two adjacent teeth is $\frac{2\pi}{22} = \frac{\pi}{11}$. The actual distance is

$$c = 6\sqrt{2 - 2\cos(\pi/11)} \approx 1.708 \text{ in.}$$

The length of the arc is

$$s = 6 \cdot \frac{\pi}{11} \approx 1.714 \text{ in.}$$

61. Solving for v_o, one finds

$$
\begin{aligned}
367 &= \frac{v_o^2}{32}\sin 86° \\
\sqrt{\frac{32(367)}{\sin 86°}} \text{ ft/sec} &= v_o \\
\sqrt{\frac{32(367)}{\sin 86°}}\frac{3600}{5280} \text{ mph} &= v_o \\
74 \text{ mph} &\approx v_o.
\end{aligned}
$$

63. Since $r = \sqrt{3^2 + 4^2} = 5$, we find

   a) $\sec\alpha = \dfrac{r}{x} = \dfrac{5}{3}$

   b) $\csc\alpha = \dfrac{r}{y} = \dfrac{5}{4}$

   c) $\cot\alpha = \dfrac{x}{y} = \dfrac{3}{4}$

65. **a)** 2   **b)** -2 **c)** $\dfrac{\sqrt{3}}{3}$

67. $-30°$

69. We begin by writing the length of a diagonal of a rectangular box. If the dimensions of a box are a-by-b-by-c, then the length of a diagonal is $\sqrt{a^2 + b^2 + c^2}$.

Suppose the ball is placed at the center of the field and 60 feet from the goal line. Then the distance between the ball and the right upright of the goal is

$$A = \sqrt{90^2 + 10^2 + 9.25^2} \approx 91.025.$$

Consider the triangle formed by the ball, and the left and right uprights of the goal. Opposite the angle θ_1 is the 10-ft horizontal bar. To simplify the calculation, we use the cosine law in Chapter 5. Then

$$18.5^2 = 2A^2 - 2A^2\cos\theta$$

and $\theta_1 \approx 11.66497°$.

Now, place the ball on the right hash mark which is 9.25 ft from the centerline. The ball is also 60 feet from the goal line. Then distance between the ball and the right upright of the goal is

$$B = \sqrt{90^2 + 10^2} \approx 90.554.$$

And, the distance between the ball and the left upright of the goal is

$$C = \sqrt{90^2 + 10^2 + 18.5^2} \approx 92.424.$$

Similarly, consider the triangle formed by the ball, and the left and right uprights of the goal. Opposite the angle θ is the 10-ft horizontal bar. Using the cosine law, we find

$$18.5^2 = B^2 + C^2 - 2BC\cos\theta_2$$

and $\theta_2 \approx 11.54654°$.

Thus, the difference between the values of θ is

$$\theta_1 - \theta_2 \approx 11.66497° - 11.54654° \approx 0.118°$$

Chapter 1 Review Exercises

1. $388° - 360° = 28°$

3. $-153°14'27'' + 359°59'60'' = 206°45'33''$

5. $180°$

7. $13\pi/5 - 2\pi = 3\pi/5 = 3 \cdot 36° = 108°$

9. $5\pi/3 = 5 \cdot 60° = 300°$

11. $270°$ **13.** $11\pi/6$ **15.** $-5\pi/3$

17.

θ deg	0	30	45	60	90	120	135	150	180
θ rad	0	$\frac{\pi}{6}$	$\frac{\pi}{4}$	$\frac{\pi}{3}$	$\frac{\pi}{2}$	$\frac{2\pi}{3}$	$\frac{3\pi}{4}$	$\frac{5\pi}{6}$	π
$\sin\theta$	0	$\frac{1}{2}$	$\frac{\sqrt{2}}{2}$	$\frac{\sqrt{3}}{2}$	1	$\frac{\sqrt{3}}{2}$	$\frac{\sqrt{2}}{2}$	$\frac{1}{2}$	0
$\cos\theta$	1	$\frac{\sqrt{3}}{2}$	$\frac{\sqrt{2}}{2}$	$\frac{1}{2}$	0	$-\frac{1}{2}$	$-\frac{\sqrt{2}}{2}$	$-\frac{\sqrt{3}}{2}$	-1

19. $-\sqrt{2}/2$ **21.** $\sqrt{3}$ **23.** $-2\sqrt{3}/3$

25. 0 **27.** 0

29. -1 **31.** $\cot(60°) = \sqrt{3}/3$ **33.** $-\sqrt{2}/2$

35. -2 **37.** $-\sqrt{3}/3$

39. 0.6947 **41.** -0.0923 **43.** 0.1869

45. $\dfrac{1}{\cos(105°4')} \approx -3.8470$

47. $\dfrac{1}{\sin(\pi/9)} \approx 2.9238$

49. $\dfrac{1}{\tan(33°44')} \approx 1.4975$

51. $45°$ **53.** $0°$

55. $30°$ **57.** $30°$

59. Note, the the length of the hypotenuse is 13.
Then $\sin(\alpha) = \text{opp/hyp} = 5/13$,
$\cos(\alpha) = \text{adj/hyp} = 12/13$,
$\tan(\alpha) = \text{opp/adj} = 5/12$,
$\csc(\alpha) = \text{hyp/opp} = 13/5$,
$\sec(\alpha) = \text{adj/hyp} = 13/12$, and
$\cot(\alpha) = \text{adj/opp} = 12/5$.

61. Form the right triangle with $a = 2$, $b = 3$.

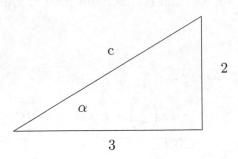

Note that $c = \sqrt{2^2 + 3^2} = \sqrt{13}$, $\tan(\alpha) = 2/3$,
so $\alpha = \tan^{-1}(2/3) \approx 33.7°$ and $\beta \approx 56.3°$.

63. Form the right triangle with $a = 3.2$ and $\alpha = 21.3°$.

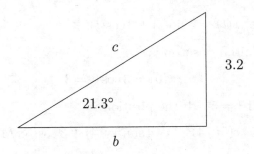

Since $\sin 21.3° = \dfrac{3.2}{c}$ and

$\tan 21.3° = \dfrac{3.2}{b}$, $c = \dfrac{3.2}{\sin 21.3°} \approx 8.8$

and $b = \dfrac{3.2}{\tan 21.3°} \approx 8.2$

Also, $\beta = 90° - 21.3° = 68.7°$

65. $\sin(\alpha) = -\sqrt{1 - \left(\dfrac{1}{5}\right)^2} = -\sqrt{\dfrac{24}{25}} = \dfrac{-2\sqrt{6}}{5}$

67. In one hour, the nozzle revolves through an angle of $\dfrac{2\pi}{8}$. The linear velocity is

$$v = r \cdot \alpha = 120 \cdot \dfrac{2\pi}{8} \approx 94.2 \text{ ft/hr}.$$

69. The height of the man is

$$s = r \cdot \alpha = 1000(0.4) \cdot \dfrac{\pi}{180} \approx 6.9813 \text{ ft}.$$

71. Form the right triangle below.

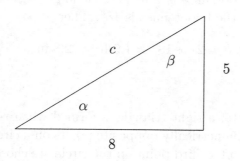

By the Pythagorean Theorem, we get

$$c = \sqrt{8^2 + 5^2} = \sqrt{89} \text{ ft.}$$

Note, $\alpha = \tan^{-1}\left(\dfrac{5}{8}\right) \approx 32.0°$ and
$\beta = 90° - \alpha \approx 58.0°$.

73. Form the right triangle below.

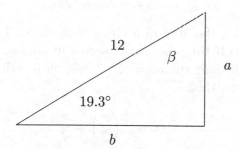

Note, $\beta = 90° - 19.3° = 70.7°$.
Also, $a = 12\sin(19.3°) \approx 4.0$ ft and
$b = 12\cos(19.3°) \approx 11.3$ ft.

75. Let s be the height of the shorter building and
let $a + b$ the height of the taller building.

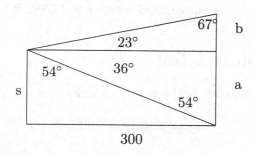

Since $\tan 54° = \dfrac{300}{s}$, get

$$s = \frac{300}{\tan 54°} \approx 218 \text{ ft.}$$

Similarly, since $\tan 36° = \dfrac{a}{300}$ and

$\tan 23° = \dfrac{b}{300}$, the height of the taller building
is

$$a + b = 300\tan 36° + 300\tan 23° \approx 345 \text{ ft.}$$

77. Let h be the height of the tower.

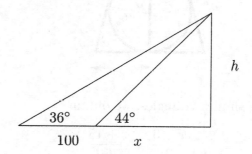

Since
$$h = (100 + x)\tan 36°$$
and
$$x = \frac{h}{\tan 44°}$$
we find

$$h = \left(100 + \frac{h}{\tan 44°}\right)\tan 36°$$

$$h = 100\tan 36° + h\frac{\tan 36°}{\tan 44°}$$

$$h = \frac{100\tan 36°}{1 - \dfrac{\tan 36°}{\tan 44°}}$$

$$h \approx 293 \text{ ft}$$

79. Draw an isosceles triangle containing the two circles as shown below. The indicated radii are perpendicular to the sides of the triangle.

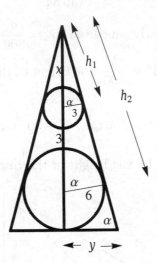

Using similar triangles, we obtain

$$\frac{x+3}{3} = \frac{x+15}{6}$$

from which we solve $x = 9$. Applying the Pythagorean theorem, we find

$$h_1 = 3\sqrt{15}, \quad h_2 = 6\sqrt{15}$$

Note, the height of the triangle is 30 ft. By similar triangles,

$$\frac{3\sqrt{15}}{3} = \frac{30}{y}$$

or $y = 2\sqrt{15}$. Thus, the base angle α of the isosceles triangle is $\alpha = \arctan\sqrt{15}$.

Then the arc lengths of the belt that wrap around the two pulleys are (using $s = r\alpha$)

$$s_1 = 6 \cdot (2\pi - 2\alpha)$$

and

$$s_2 = 3 \cdot 2\alpha = 6\alpha.$$

Hence, the length of the belt is

$$s_1 + s_2 + 2(h_2 - h_1) \ =$$

$$12\pi - 6\arctan(\sqrt{15}) + 6\sqrt{15} \ \approx$$

$$53.0 \text{ in.}$$

81. Consider a right triangle whose height is h, the hypotenuse is 2 ft, and the angle between h and the hypotenuse is $18°$. Then

$$h = 2\cos 18° \approx 1.9 \text{ ft} = 22.8 \text{ in.}$$

83.

a) Note, a right triangle is formed when two diametrically oppposite points on a circle and a third point on the circle are chosen as vertices of a triangle.

Then the angle spanned by the sector is

$$\theta = \cos^{-1}\frac{x}{2}.$$

Since the area of the sector is $A = \dfrac{\theta r^2}{2}$, we find

$$A = \frac{x^2}{2}\cos^{-1}\left(\frac{x}{2}\right).$$

b) Note, the area of a circle with radius 1 is π. If the area of the sector in part a) is one-half the area of a circle with radius one, then

$$\frac{x^2}{2}\cos^{-1}\left(\frac{x}{2}\right) = \frac{\pi}{2}.$$

Thus, the radius of the blade is either $x = \sqrt{2}$ ft or $x = \sqrt{3}$ ft.

c) Using a graphing calculator, we find that the area

$$A = \frac{x^2}{2}\cos^{-1}\left(\frac{x}{2}\right)$$

is maximized when $x \approx 1.5882$ ft.

Chapter 1 Test

1. Since $60°$ is the reference angle, we get

$$\cos 420° = \cos(60°) = 1/2.$$

2. Since $30°$ is the reference angle, we get

$$\sin(-390°) = -\sin(30°) = -\frac{1}{2}.$$

3. $\dfrac{\sqrt{2}}{2}$ **4.** $\dfrac{1}{2}$ **5.** $\dfrac{\sqrt{3}}{3}$ **6.** $\sqrt{3}$

7. Undefined, since $\dfrac{1}{\cos(\pi/2)} = \dfrac{1}{0}$

8. $\dfrac{1}{\sin(-\pi/2)} = \dfrac{1}{-1} = -1$

9. Undefined, since $(-1, 0)$ lies on the terminal side of angle -3π and

$$\cot(-3\pi) = \frac{x}{y} = \frac{-1}{0}.$$

10. Since $(-1, -1)$ lies on the terminal side of angle $225°$, we get

$$\cot(225°) = \frac{x}{y} = \frac{-1}{-1} = 1.$$

11. $\dfrac{\pi}{4}$ or $45°$ **12.** $\dfrac{\pi}{6}$ or $30°$

13. Since $46°24'6'' \approx 0.8098619$, the arclength is $s = r\alpha = 35.62(0.8098619) \approx 28.85$ meters.

14. $2.34 \cdot \dfrac{180°}{\pi} \approx 134.07°$

15. Coterminal since $2200° - 40° = 2160° = 6 \cdot 360°$.

16. $\cos(\alpha) = -\sqrt{1 - \left(\dfrac{1}{4}\right)^2} = -\dfrac{\sqrt{15}}{4}$

17. $\omega = 103 \cdot 2\pi \approx 647.2$ radians/minute

18. In one minute, the wheel turns through an arclength of $13(103 \cdot 2\pi)$ inches.

Multiplying this by $\dfrac{60}{12 \cdot 5280}$ results in the speed in mph which is 7.97 mph.

19. Since $r = \sqrt{x^2 + y^2} = \sqrt{5^2 + (-2)^2} = \sqrt{29}$,

we find $\sin\alpha = \dfrac{y}{r} = \dfrac{-2}{\sqrt{29}} = \dfrac{-2\sqrt{29}}{29}$,

$\cos\alpha = \dfrac{x}{r} = \dfrac{5}{\sqrt{29}} = \dfrac{5\sqrt{29}}{29}$,

$\tan\alpha = \dfrac{y}{x} = \dfrac{-2}{5}$,

$$\csc\alpha = \frac{r}{y} = \frac{\sqrt{29}}{-2} = -\frac{\sqrt{29}}{2},$$

$$\sec\alpha = \frac{r}{x} = \frac{\sqrt{29}}{5}, \text{ and}$$

$$\cot\alpha = \frac{x}{y} = \frac{5}{-2} = -\frac{5}{2}.$$

20. Consider the right triangle below.

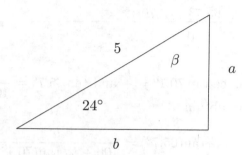

Note, $\beta = 90° - 24° = 66°$.
Then $a = 5\sin(24°) \approx 2.0$ ft and $b = 5\cos(24°) \approx 4.6$ ft.

21. Let h be the height of the head.

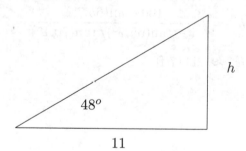

Since

$$\tan(48°) = \frac{h}{11}$$

we find

$$h = 11\tan 48° \approx 12.2 \text{ m}.$$

22. Let h be the height of the building.

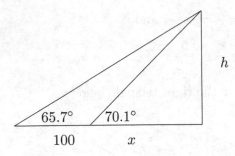

Since $\tan 70.1^o = \dfrac{h}{x}$ and $\tan 65.7^\circ = \dfrac{h}{100+x}$, we obtain

$$\tan(65.7^\circ) = \frac{h}{100 + h/\tan(70.1^\circ)}$$

$$100 \cdot \tan(65.7^\circ) + h \cdot \frac{\tan(65.7^o)}{\tan(70.1^\circ)} = h$$

$$100 \cdot \tan(65.7^\circ) = h\left(1 - \frac{\tan(65.7^\circ)}{\tan(70.1^\circ)}\right)$$

$$h \;=\; \frac{100 \cdot \tan(65.7^\circ)}{1 - \tan(65.7^\circ)/\tan(70.1^\circ)}$$

$$h \;\approx\; 1117 \text{ ft.}$$

For Thought

1. True

2. False, since the range of $y = 4\sin(x)$ is $[-4, 4]$, we get that the range of $y = 4\sin(x) + 3$ is $[-4 + 3, 4 + 3]$ or $[-1, 7]$.

3. True, since the range of $y = \cos(x)$ is $[-1, 1]$, we find that the range of $y = \cos(x) - 5$ is $[-1 - 5, 1 - 5]$ or $[-6, -4]$.

4. False, the phase shift is $-\pi/6$.

5. False, the graph of $y = \sin(x + \pi/6)$ lies $\pi/6$ to the *left* of the graph of $y = \sin(x)$.

6. True, since $\cos(5\pi/6 - \pi/3) = \cos(\pi/2) = 0$ and $\cos(11\pi/6 - \pi/3) = \cos(3\pi/2) = 0$.

7. False, for if $x = \pi/2$ we find $\sin(\pi/2) = 1 \neq \cos(\pi/2 + \pi/2) = \cos(\pi) = -1$.

8. False, the minimum value is -3.

9. True, since the maximum value of $y = -2\cos(x)$ is 2, we get that the maximum value of $y = -2\cos(x) + 4$ is $2 + 4$ or 6.

10. True, since $(-\pi/6 + \pi/3, 0) = (\pi/6, 0)$.

2.1 Exercises

1. $\sin\alpha$, $\cos\alpha$

3. sine wave

5. fundamental cycle

7. phase shift

9. starting, maximum, inflection, minimum, ending

11. 0

13. $\dfrac{\sin(\pi/3)}{\cos(\pi/3)} = \dfrac{\sqrt{3}/2}{1/2} = \sqrt{3}$.

15. 1/2

17. $\dfrac{1}{\cos(\pi/3)} = \dfrac{1}{1/2} = 2$.

19. 0

21. 0

23. -2

25. $-\dfrac{\sqrt{2}}{2}$

27. $(0 + \pi/4, 0) = (\pi/4, 0)$

29. $(\pi/2 + \pi/4, 3) = (3\pi/4, 3)$

31. $(-\pi/2 + \pi/4, -1) = (-\pi/4, -1)$

33. $(\pi + \pi/4, 0) = (5\pi/4, 0)$

35. $(\pi/3 - \pi/3, 0) = (0, 0)$

37. $(\pi - \pi/3, 1) = (2\pi/3, 1)$

39. $(\pi/2 - \pi/3, -1) = (\pi/6, -1)$

41. $(-\pi - \pi/3, 1) = (-4\pi/3, 1)$

43. $(\pi + \pi/6, -1 + 2) = (7\pi/6, 1)$

45. $(\pi/2 + \pi/6, 0 + 2) = (2\pi/3, 2)$

47. $(-3\pi/2 + \pi/6, 1 + 2) = (-4\pi/3, 3)$

49. $(2\pi + \pi/6, -4 + 2) = (13\pi/6, -2)$

51. $\left(\dfrac{\pi + 2\pi}{2}, 0\right) = \left(\dfrac{3\pi}{2}, 0\right)$

53. $\left(\dfrac{0 + \pi/4}{2}, 2\right) = \left(\dfrac{\pi}{8}, 2\right)$

55. $\left(\dfrac{\pi/6 + \pi/2}{2}, 1\right) = \left(\dfrac{\pi}{3}, 1\right)$

57. $\left(\dfrac{\pi/3 + \pi/2}{2}, -4\right) = \left(\dfrac{5\pi}{12}, -4\right)$

59. $P(0, 0)$, $Q(\pi/4, 2)$, $R(\pi/2, 0)$, $S(3\pi/4, -2)$

61. $P(\pi/4, 0)$, $Q(5\pi/8, 2)$, $R(\pi, 0)$, $S(11\pi/8, -2)$

63. $P(0, 2)$, $Q(\pi/12, 3)$, $R(\pi/6, 2)$, $S(\pi/4, 1)$

65. Amplitude 2, period 2π, phase shift 0, range $[-2, 2]$

67. Amplitude 1, period 2π, phase shift $\pi/2$, range $[-1, 1]$

69. Amplitude 2, period 2π, phase shift $-\pi/3$, range $[-2, 2]$

71. Amplitude 1, phase shift 0, range $[-1, 1]$, some points are $(0,0)$, $\left(\dfrac{\pi}{2}, -1\right)$,

$(\pi, 0)$, $\left(\dfrac{3\pi}{2}, 1\right)$, $(2\pi, 0)$

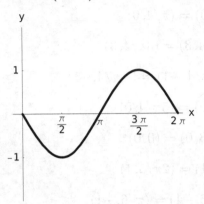

73. Amplitude 3, phase shift 0, range $[-3, 3]$, some points are $(0,0)$, $\left(\dfrac{\pi}{2}, -3\right)$, $(\pi, 0)$, $\left(\dfrac{3\pi}{2}, 3\right)$, $(2\pi, 0)$

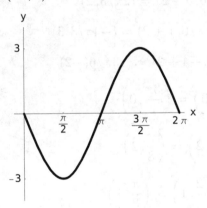

75. Amplitude 1/2, phase shift 0, range $[-1/2, 1/2]$, some points are $(0, 1/2)$, $(\pi/2, 0)$, $(\pi, -1/2)$ $(3\pi/2, 0)$, $(2\pi, 1/2)$

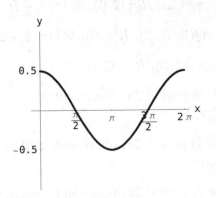

77. Amplitude 1, phase shift $-\pi$, range $[-1, 1]$, some points are $(-\pi, 0)$, $(-\pi/2, 1)$, $(0, 0)$, $\left(\dfrac{\pi}{2}, -1\right)$, $(\pi, 0)$

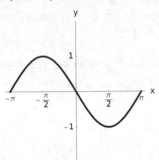

79. Amplitude 1, phase shift $\pi/3$, range $[-1, 1]$, some points are $\left(\dfrac{\pi}{3}, 1\right)$, $\left(\dfrac{5\pi}{6}, 0\right)$, $\left(\dfrac{4\pi}{3}, -1\right)$, $\left(\dfrac{11\pi}{6}, 0\right)$, $\left(\dfrac{7\pi}{3}, 1\right)$,

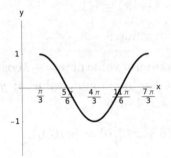

81. Amplitude 1, phase shift 0, range $[1, 3]$, some points are $(0, 3)$, $\left(\dfrac{\pi}{2}, 2\right)$, $(\pi, 1)$ $\left(\dfrac{3\pi}{2}, 2\right)$,

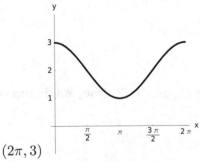

$(2\pi, 3)$

83. Amplitude 1, phase shift 0, range $[-2, 0]$, some points are $(0, -1)$, $\left(\frac{\pi}{2}, -2\right)$, $(\pi, -1)$, $\left(\frac{3\pi}{2}, 0\right)$, $(2\pi, -1)$

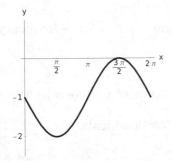

85. Amplitude 1, phase shift $-\pi/4$, range $[1, 3]$, some points are $\left(-\frac{\pi}{4}, 2\right)$, $\left(\frac{\pi}{4}, 3\right)$, $\left(\frac{3\pi}{4}, 2\right)$, $\left(\frac{5\pi}{4}, 1\right)$, $\left(\frac{7\pi}{4}, 2\right)$

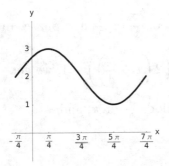

87. Amplitude 2, phase shift $-\pi/6$, range $[-1, 3]$, some points are $\left(-\frac{\pi}{6}, 3\right)$, $\left(\frac{\pi}{3}, 1\right)$, $\left(\frac{5\pi}{6}, -1\right)$,

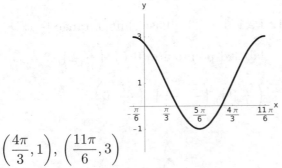

$\left(\frac{4\pi}{3}, 1\right)$, $\left(\frac{11\pi}{6}, 3\right)$

89. Amplitude 2, phase shift $\pi/3$, range $[-1, 3]$, some points are $\left(\frac{\pi}{3}, 1\right)$, $\left(\frac{5\pi}{6}, -1\right)$, $\left(\frac{4\pi}{3}, 1\right)$, $\left(\frac{11\pi}{6}, 3\right)$ $\left(\frac{7\pi}{3}, 1\right)$

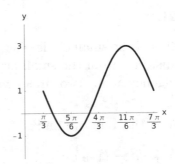

91. Note the amplitude is 2, phase shift is $\pi/2$, and the period is 2π. Then $A = 2$, $C = \pi/2$, and $B = 1$. An equation is $y = 2\sin\left(x - \frac{\pi}{2}\right)$.

93. Note the amplitude is 2, phase shift $-\pi$, the period is 2π, and the vertical shift is 1 unit up. Then $A = 2$, $C = -\pi$, $B = 1$, and $D = 1$.

An equation is $y = 2\sin(x + \pi) + 1$.

95. Note the amplitude is 2, phase shift is $\pi/2$, and the period is 2π. Then $A = 2$, $C = \pi/2$, and $B = 1$. An equation is $y = 2\cos\left(x - \frac{\pi}{2}\right)$.

97. Note, the amplitude is 2, phase shift is π, the period is 2π, and the vertical shift is 1 unit up. Then $A = 2$, $C = \pi$, $B = 1$, and $D = 1$.

An equation is $y = 2\cos(x - \pi) + 1$.

99. $y = \sin\left(x - \frac{\pi}{4}\right)$

101. $y = \sin\left(x + \frac{\pi}{2}\right)$

103. $y = -\cos\left(x - \frac{\pi}{5}\right)$

105. $y = -\cos\left(x - \frac{\pi}{8}\right) + 2$

107. $y = -3\cos\left(x + \frac{\pi}{4}\right) - 5$

109. A determines the stretching, shrinking, or reflection about the x-axis, C is the phase shift, and D is the vertical translation.

111. $\pi/6$

113. $\dfrac{93 \cdot 10^6 \cdot 2\pi}{365(24)} \approx 67,000$ mph

115. $\arcsin(0.36) \approx 21.1°$

117. Let r be the radius of the small circle, and let x be the closest distance from the small circle to the point of tangency of any two circles with radius 1.

By the Pythagorean theorem, we find

$$1 + (x+r)^2 = (1+r)^2$$

and

$$1 + (1 + 2r + x)^2 = 2^2.$$

The second equation may be written as

$$1 + (r+1)^2 + 2(r+1)(r+x) + (r+x)^2 = 4.$$

Using the first equation, the above equation simplifies to

$$(r+1)^2 + 2(r+1)(r+x) + (1+r)^2 = 4$$

or

$$(r+1)^2 + (r+1)(r+x) = 2.$$

Since (from first equation, again)

$$x + r = \sqrt{(1+r)^2 - 1}$$

we obtain

$$(r+1)^2 + (r+1)\left(\sqrt{(1+r)^2 - 1}\right) = 2.$$

Solving for r, we find

$$r = \frac{2\sqrt{3} - 3}{3}.$$

For Thought

1. True, since $B = 4$ and the period is $\frac{2\pi}{B} = \frac{\pi}{2}$.

2. False, since $B = 2\pi$ and the period is $\frac{2\pi}{B} = 1$.

3. True, since $B = \pi$ and the period is $\frac{2\pi}{B} = 2$.

4. True, since $B = 0.1\pi$ and the period is $\frac{2\pi}{0.1\pi} = 20$.

5. False, the phase shift is $-\frac{\pi}{12}$.

6. False, the phase shift is $-\frac{\pi}{8}$.

7. True, since the period is $P = 2\pi$ the frequency is $\frac{1}{P} = \frac{1}{2\pi}$.

8. True, since the period is $P = 2$ the frequency is $\frac{1}{P} = \frac{1}{2}$.

9. False, rather the graphs of $y = \cos(x)$ and $y = \sin\left(x + \frac{\pi}{2}\right)$ are identical.

10. True

2.2 Exercises

1. period

3. amplitude

5. Amplitude 3, period $\frac{2\pi}{4}$ or $\frac{\pi}{2}$, and phase shift 0

7. Since $y = -2\cos\left(2\left(x + \frac{\pi}{4}\right)\right) - 1$, we get amplitude 2, period $\frac{2\pi}{2}$ or π, and phase shift $-\frac{\pi}{4}$.

9. Since $y = -2\sin(\pi(x-1))$, we get amplitude 2, period $\frac{2\pi}{\pi}$ or 2, and phase shift 1.

11. Period $2\pi/3$, phase shift 0, range $[-1, 1]$, labeled points are $(0,0)$, $\left(\frac{\pi}{6}, 1\right)$, $\left(\frac{\pi}{3}, 0\right)$, $\left(\frac{\pi}{2}, -1\right)$, $\left(\frac{2\pi}{3}, 0\right)$

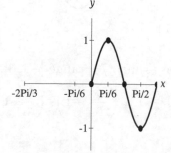

13. Period π, phase shift 0, range $[-1, 1]$, labeled points are $(0, 0)$, $\left(\dfrac{\pi}{4}, -1\right)$, $\left(\dfrac{\pi}{2}, 0\right)$, $\left(\dfrac{3\pi}{4}, 1\right)$, $(\pi, 0)$

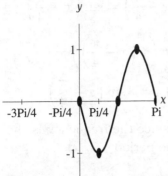

15. Period $\pi/2$, phase shift 0, range $[1, 3]$, labeled points are $(0, 3)$, $\left(\dfrac{\pi}{8}, 2\right)$, $\left(\dfrac{\pi}{4}, 1\right)$, $\left(\dfrac{3\pi}{8}, 2\right)$, $\left(\dfrac{\pi}{2}, 3\right)$

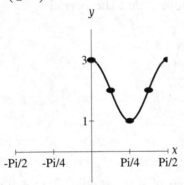

17. Period 8π, phase shift 0, range $[1, 3]$, labeled points are $(0, 2)$, $(2\pi, 1)$, $(4\pi, 2)$, $(6\pi, 3)$, $(8\pi, 2)$

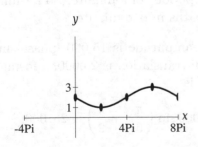

19. Period 6, phase shift 0, range $[-1, 1]$, labeled points are $(0, 0)$, $(1.5, 1)$, $(3, 0)$, $(4.5, -1)$, $(6, 0)$

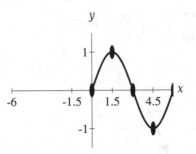

21. Period π, phase shift $\pi/2$, range $[-1, 1]$, labeled points are $\left(\dfrac{\pi}{2}, 0\right)$, $\left(\dfrac{3\pi}{4}, 1\right)$, $(\pi, 0)$, $\left(\dfrac{5\pi}{4}, -1\right)$, $\left(\dfrac{3\pi}{2}, 0\right)$

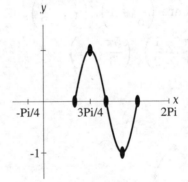

23. Period 4, phase shift -3, range $[-1, 1]$, labeled points are $(-3, 0)$, $(-2, 1)$, $(-1, 0)$, $(0, -1)$, $(1, 0)$

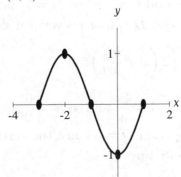

25. Period π, phase shift $-\pi/6$, range $[-1,3]$, labeled points are $\left(-\dfrac{\pi}{6},3\right)$, $\left(\dfrac{\pi}{12},1\right)$,

$\left(\dfrac{\pi}{3},-1\right)$, $\left(\dfrac{7\pi}{12},1\right)$, $\left(\dfrac{5\pi}{6},3\right)$

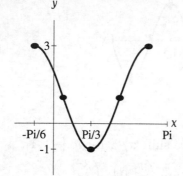

27. Period $\dfrac{2\pi}{3}$, phase shift $\dfrac{\pi}{6}$, range $\left[-\dfrac{3}{2},-\dfrac{1}{2}\right]$,

labeled points are $\left(\dfrac{\pi}{6},-1\right)$, $\left(\dfrac{\pi}{3},-\dfrac{3}{2}\right)$,

$\left(\dfrac{\pi}{2},-1\right)$, $\left(\dfrac{2\pi}{3},-\dfrac{1}{2}\right)$, $\left(\dfrac{5\pi}{6},-1\right)$

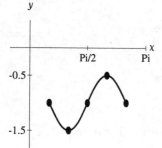

29. Note, $A=2$, period is π and so $B=2$, phase shift is $C=\dfrac{\pi}{4}$, and $D=0$ or no vertical shift.

Then $y = 2\sin\left(2\left(x-\dfrac{\pi}{4}\right)\right)$.

31. Note, $A=3$, period is $\dfrac{4\pi}{3}$ and so $B=\dfrac{3}{2}$, phase shift is $C=-\dfrac{\pi}{3}$, and $D=3$ since the vertical shift is three units up.

Then $y = 3\sin\left(\dfrac{3}{2}\left(x+\dfrac{\pi}{3}\right)\right)+3$.

33. $\dfrac{\pi}{4}-\dfrac{\pi}{4}=0$

35. $f\left(g\left(\dfrac{\pi}{4}\right)\right) = f(0) = \sin(0) = 0$

37. $h\left(f\left(g\left(\dfrac{\pi}{4}\right)\right)\right) = h\left(f\left(0\right)\right) = h\left(\sin\left(0\right)\right) = $
$h(0) = 3\cdot 0 = 0$

39. $f\left(g\left(x\right)\right) = f\left(x-\dfrac{\pi}{4}\right) = \sin\left(x-\dfrac{\pi}{4}\right)$

41. $h\left(f\left(g\left(x\right)\right)\right) = h\left(f\left(x-\dfrac{\pi}{4}\right)\right) = $
$h\left(\sin\left(x-\dfrac{\pi}{4}\right)\right) = 3\sin\left(x-\dfrac{\pi}{4}\right)$

43. 100 cycles/sec since the frequency is the reciprocal of the period

45. Frequency is $\dfrac{1}{0.025} = 40$ cycles per hour

47. Substitute $v_o = 6$, $\omega = 2$, and $x_o = 0$ into $x(t) = \dfrac{v_o}{\omega}\cdot\sin(\omega t) + x_o\cdot\cos(\omega t)$.

Then $x(t) = 3\sin(2t)$.
The amplitude is 3 and the period is π.

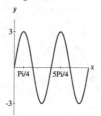

49. 11 years

51. Note, the range of $v = 400\sin(60\pi t)+900$ is $[-400+900,400+900]$ or $[500,1300]$.
(a) Maximum volume is 1300 cc and mininum volume is 500 cc
(b) The runner takes a breath every 1/30 (which is the period) of a minute. So a runner makes 30 breaths in one minute.

53. Period is 12, amplitude is 15,000, phase-shift is -3, vertical translation is 25,000, a formula for the curve is

$$y = 15,000\sin\left(\dfrac{\pi}{6}x+\dfrac{\pi}{2}\right)+25,000;$$

for April (when $x=4$), the revenue is

$$15,000\sin\left(\dfrac{\pi}{6}x+\dfrac{\pi}{2}\right)+25,000 \approx \$17,500.$$

55.

a) period is 40, amplitude is 65, an equation for the sine wave is

$$y = 65 \sin\left(\frac{\pi}{20}x\right)$$

b) 40 days

c) $65 \sin\left(\frac{\pi}{20}(36)\right) \approx -38.2$ meters/second

d) The new planet is between Earth and Rho.

57. Since the period is $20 = \frac{2\pi}{B}$, we get $B = \frac{\pi}{10}$.

Also, the amplitude is 1 and the vertical translation is 1. An equation for the swell is

$$y = \sin\left(\frac{\pi}{10}x\right) + 1.$$

59. The sine regression curve is

$$y = 50 \sin(0.214x - 0.615) + 48.8$$

or approximately

$$y = 50 \sin(0.21x - 0.62) + 48.8$$

The period is

$$\frac{2\pi}{b} = \frac{2\pi}{0.214} \approx 29.4 \text{ days.}$$

When $x = 39$, we find

$$y = 50 \sin(0.214(39) - 0.615) + 48.8 \approx 98\%$$

On February 8, 2020, 98% of the moon is illuminated. Shown below is a graph of the regression equation and the data points.

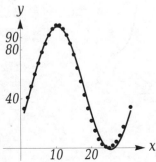

63. Amplitude $A = \frac{1}{2}$,

period $B = \frac{2\pi}{B} = \frac{2\pi}{1} = 2\pi$,

phase shift $\frac{\pi}{2}$,

period is $[-\frac{1}{2} + 3, \frac{1}{2} + 3]$ or $[2.5, 3.5]$

65. If x is the height of the tree, then $\tan 30° = h/500$ or

$$h = 500 \tan 30° \approx 289 \text{ ft.}$$

67. Note, the sum of the two angles is $32°37' + 48°39' = 80°76' = 81°16'$.

The third angle is $179°60' - 81°16' = 98°44'$.

69. One possibility is

$$\text{WRONG} = 25938$$

and

$$\text{RIGHT} = 51876.$$

For Thought

1. True, since $\sec(\pi/4) = \dfrac{1}{\cos(\pi/4)} = \dfrac{1}{\sin(\pi/4)}$.

2. True, since $\csc(x) = \dfrac{1}{\sin(x)}$.

3. True, since $\csc(\pi/2) = \dfrac{1}{\sin(\pi/2)} = \dfrac{1}{1} = 1$.

4. False, since $\dfrac{1}{\cos(\pi/2)}$ or $\dfrac{1}{0}$ is undefined.

5. True, since $B = 2$ and $\dfrac{2\pi}{B} = \dfrac{2\pi}{2} = \pi$.

6. True, since $B = \pi$ and $\dfrac{2\pi}{B} = \dfrac{2\pi}{\pi} = 2$.

7. False, rather the graphs of $y = 2\csc x$ and

$$y = \frac{2}{\sin x} \text{ are identical.}$$

8. True, since the maximum and minimum of $0.5 \csc x$ are ± 0.5.

9. True, since $\frac{\pi}{2} + k\pi$ are the zeros of $y = \cos x$
we get that the asymptotes of $y = \sec(2x)$ are
$2x = \frac{\pi}{2} + k\pi$ or $x = \frac{\pi}{4} + \frac{k\pi}{2}$. If $k = \pm 1$,
we get the asymptotes $x = \pm\frac{\pi}{4}$.

10. True, since if we substitute $x = 0$ in $\dfrac{1}{\csc(4x)}$
we get $\dfrac{1}{\csc(0)}$ or $\dfrac{1}{0}$ which is undefined.

2.3 Exercises

1. domain

3. asymptote

5. $\dfrac{1}{\cos(\pi/3)} = \dfrac{1}{1/2} = 2$

7. $\dfrac{1}{\sin(-\pi/4)} = \dfrac{1}{-1/\sqrt{2}} = -\sqrt{2}$

9. Undefined, since $\dfrac{1}{\cos(\pi/2)} = \dfrac{1}{0}$

11. Undefined, since $\dfrac{1}{\sin(\pi)} = \dfrac{1}{0}$

13. $\dfrac{1}{\cos 1.56} \approx 92.6$

15. $\dfrac{1}{\sin 0.01} \approx 100.0$

17. $\dfrac{1}{\sin 3.14} \approx 627.9$

19. $\dfrac{1}{\cos 4.71} \approx -418.6$

21. Since $B = 2$, the period is $\dfrac{2\pi}{B} = \dfrac{2\pi}{2}$ or π

23. Since $B = \dfrac{3}{2}$, the period is $\dfrac{2\pi}{B} = \dfrac{2\pi}{3/2}$ or $\dfrac{4\pi}{3}$

25. Since $B = \pi$, the period is $\dfrac{2\pi}{B} = \dfrac{2\pi}{\pi}$ or 2

27. $(-\infty, -2] \cup [2, \infty)$

29. $(-\infty, -1/2] \cup [1/2, \infty)$

31. Since the range of $y = \sec(\pi x - 3\pi)$ is

$$(-\infty, -1] \cup [1, \infty),$$

the range of $y = \sec(\pi x - 3\pi) - 1$ is

$$(-\infty, -1 - 1] \cup [1 - 1, \infty)$$

or equivalently

$$(-\infty, -2] \cup [0, \infty).$$

33. period 2π, asymptotes $x = \dfrac{\pi}{2} + k\pi$,
range $(-\infty, -2] \cup [2, \infty)$

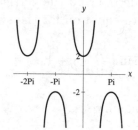

35. period $2\pi/3$, asymptotes $3x = \dfrac{\pi}{2} + k\pi$ or
$x = \dfrac{\pi}{6} + \dfrac{k\pi}{3}$, range $(-\infty, -1] \cup [1, \infty)$

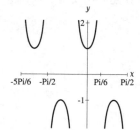

37. period $\dfrac{2\pi}{1} = 2\pi$,asymptotes $x + \dfrac{\pi}{4} = \dfrac{\pi}{2} + k\pi$
or $x = \dfrac{\pi}{4} + k\pi$, range $(-\infty, -1] \cup [1, \infty)$

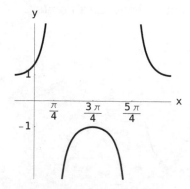

39. period $\dfrac{2\pi}{1/2} = 4\pi$, asymptotes $\dfrac{x}{2} = \dfrac{\pi}{2} + k\pi$ or

$x = \pi + 2k\pi$, range $(-\infty, -1] \cup [1, \infty)$

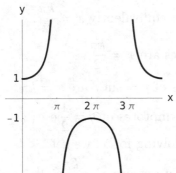

41. period $\dfrac{2\pi}{\pi} = 2$, asymptotes $\pi x = \dfrac{\pi}{2} + k\pi$ or

$x = \dfrac{1}{2} + k$, range $(-\infty, -2] \cup [2, \infty)$

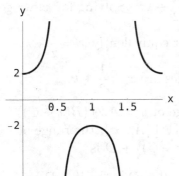

43. period $\dfrac{2\pi}{2} = \pi$, asymptotes $2x = \dfrac{\pi}{2} + k\pi$ or

$x = \dfrac{\pi}{4} + \dfrac{k\pi}{2}$, and since the range of

$y = 2\sec(2x)$ is $(-\infty, -2] \cup [2, \infty)$ then

the range of $y = 2 + 2\sec(2x)$ is

$(-\infty, -2 + 2] \cup [2 + 2, \infty)$ or $(-\infty, 0] \cup [4, \infty)$.

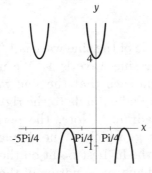

45. period 2π, asymptotes $x = k\pi$,

range $(-\infty, -2] \cup [2, \infty)$

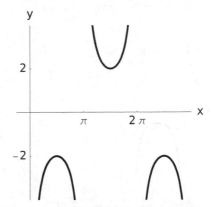

47. period 2π, asymptotes $x + \dfrac{\pi}{2} = k\pi$ or

$x = -\dfrac{\pi}{2} + k\pi$ or $x = \dfrac{\pi}{2} + k\pi$,

range $(-\infty, -1] \cup [1, \infty)$

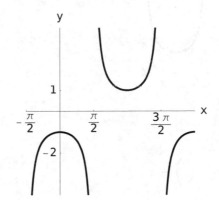

49. period $\dfrac{2\pi}{2}$ or π, asymptotes $2x - \dfrac{\pi}{2} = k\pi$

or $x = \dfrac{\pi}{4} + \dfrac{k\pi}{2}$, range $(-\infty, -1] \cup [1, \infty)$

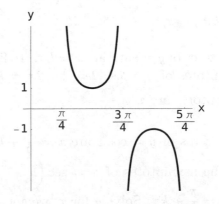

51. period $\dfrac{2\pi}{\pi/2}$ or 4, asymptotes $\dfrac{\pi x}{2} - \dfrac{\pi}{2} = k\pi$ or

$x = 1 + 2k$, range $(-\infty, -1] \cup [1, \infty)$

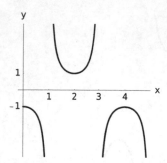

53. period $\dfrac{2\pi}{\pi/2}$ or 4, asymptotes $\dfrac{\pi x}{2} + \dfrac{\pi}{2} = k\pi$

or $x = -1 + 2k$ or $x = 1 + 2k$,

range $(-\infty, -1] \cup [1, \infty)$

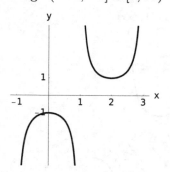

55. $y = \sec\left(x - \dfrac{\pi}{2}\right) + 1$

57. $y = -\csc\left(x + 1\right) + 4$

59. Since the zeros of $y = \cos x$ are $x = \dfrac{\pi}{2} + k\pi$,

the vertical asymptotes of $y = \sec x$ are

$x = \dfrac{\pi}{2} + k\pi$.

61. Note, the zeros of $y = \sin x$ are $x = k\pi$. To find
the asymptotes of $y = \csc(2x)$, let $2x = k\pi$.
The asymptotes are $x = \dfrac{k\pi}{2}$.

63. Note, the zeros of $y = \cos x$ are $x = \dfrac{\pi}{2} + k\pi$.

To find the asymptotes of $y = \sec\left(x - \dfrac{\pi}{2}\right)$,

let $x - \dfrac{\pi}{2} = \dfrac{\pi}{2} + k\pi$. Solving for x, we get

$x = \pi + k\pi$ or equivalently $x = k\pi$.
The asymptotes are $x = k\pi$.

65. Note, the zeros of $y = \sin x$ are $x = k\pi$.
To find the asymptotes of $y = \csc\left(2x - \pi\right)$,
let $2x - \pi = k\pi$. Solving for x, we get

$x = \dfrac{\pi}{2} + \dfrac{k\pi}{2}$ or equivalently $x = \dfrac{k\pi}{2}$.

The asymptotes are $x = \dfrac{k\pi}{2}$.

67. Note, the zeros of $y = \sin x$ are $x = k\pi$.

To find the asymptotes of $y = \dfrac{1}{2}\csc\left(2x\right) + 4$,

let $2x = k\pi$. Solving for x, we get

that the asymptotes are $x = \dfrac{k\pi}{2}$.

69. Note, the zeros of $y = \cos x$ are $x = \dfrac{\pi}{2} + k\pi$.

To find the asymptotes of $y = \sec\left(\pi x + \pi\right)$,

let $\pi x + \pi = \dfrac{\pi}{2} + k\pi$. Solving for x, we get

$x = -\dfrac{1}{2} + k$ or equivalently $x = \dfrac{1}{2} + k$.

The asymptotes are $x = \dfrac{1}{2} + k$.

71. Since the range of $y = A\sec(B(x - C))$
is $(-\infty, -|A|\,] \cup [\,|A|, \infty)$, the range of
$y = A\sec(B(x - C)) + D$ is

$$(-\infty, -|A| + D\,] \cup [\,|A| + D, \infty).$$

73. $\sin\alpha = y$, $\cos\alpha = x$

75. Note $f(x) = 5\cos\left(2\left(x - \dfrac{\pi}{2}\right)\right) + 3$.

The amplitude is $A = 5$,

period is $\dfrac{2\pi}{B} = \dfrac{2\pi}{2} = \pi$, phase shift is $C = \dfrac{\pi}{2}$,

and the range is $[-5 + 3, 5 + 3] = [-2, 8]$.

77. $\beta = 0$

79. The amplitude of the sine wave is 1/2 since the
height of the sine wave is 1. We use a coordinate system such that the sine wave begins
at the origin and extends to the right side and
the first quadrant. Note, the period of the
sine wave is π, which is the diameter of the
tube. Then the highest point on the sine wave
is $(\pi/2, 1)$. Thus, an equation of the sine wave
is

$$y = -\dfrac{1}{2}\cos(2x) + \dfrac{1}{2}.$$

For Thought

1. True, since $\tan x = \dfrac{\sin x}{\cos x}$.

2. True, since $\cot x = \dfrac{1}{\tan x}$ provided $\tan x \neq 0$.

3. False, since $\cot\left(\dfrac{\pi}{2}\right) = 0$ and $\tan\left(\dfrac{\pi}{2}\right)$ is undefined.

4. True, since $\dfrac{\sin 0}{\cos 0} = \dfrac{0}{1} = 0$.

5. False, since $\dfrac{\sin(\pi/2)}{\cos(\pi/2)}$ or $\dfrac{1}{0}$ is undefined.

6. False, since $\dfrac{\sin(5\pi/2)}{\cos(5\pi/2)}$ or $\dfrac{1}{0}$ is undefined.

7. True

8. False, the range of $y = \cot x$ is $(-\infty, \infty)$.

9. True, since $\tan\left(3 \cdot \left(\pm\dfrac{\pi}{6}\right)\right) = \tan\left(\pm\dfrac{\pi}{2}\right)$
or $\dfrac{\pm 1}{0}$ is undefined.

10. True, since $\cot\left(4 \cdot \left(\pm\dfrac{\pi}{4}\right)\right) = \cot\left(\pm\pi\right)$
or $\dfrac{\pm 1}{0}$ is undefined.

2.4 Exercises

1. tangent

3. domain

5. inflection

7. $\dfrac{\sin(\pi/3)}{\cos(\pi/3)} = \dfrac{\sqrt{3}/2}{1/2} = \dfrac{\sqrt{3}}{2} \cdot \dfrac{2}{1} = \sqrt{3}$

9. Undefined, since $\dfrac{\sin(\pi/2)}{\cos(\pi/2)}$ has the form $\dfrac{1}{0}$

11. $\dfrac{\sin(\pi)}{\cos(\pi)} = \dfrac{0}{-1} = 0$

13. $\dfrac{\cos(\pi/4)}{\sin(\pi/4)} = \dfrac{\sqrt{2}/2}{\sqrt{2}/2} = 1$

15. Undefined, since $\dfrac{\cos(0)}{\sin(0)}$ has the form $\dfrac{1}{0}$

17. $\dfrac{\cos(\pi/2)}{\sin(\pi/2)} = \dfrac{0}{1} = 0$

19. 92.6

21. -108.6

23. $\dfrac{1}{\tan 0.002} \approx 500.0$

25. $\dfrac{1}{\tan(-0.002)} \approx -500.0$

27. Since $B = 8$, the period is $\dfrac{\pi}{B} = \dfrac{\pi}{8}$.

29. Since $B = \pi$, the period is $\dfrac{\pi}{B} = \dfrac{\pi}{\pi} = 1$.

31. Since $B = \dfrac{\pi}{3}$, the period is $\dfrac{\pi}{B} = \dfrac{\pi}{\pi/3} = 3$.

33. Since $B = 3$, the period is $\dfrac{\pi}{B} = \dfrac{\pi}{3}$.

35. $y = \tan(3x)$ has period $\dfrac{\pi}{3}$, and if $3x = \dfrac{\pi}{2} + k\pi$
then the asymptotes are $x = \dfrac{\pi}{6} + \dfrac{k\pi}{3}$ for any integer k.

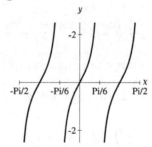

37. $y = \tan(\pi x)$ has period 1, and if $\pi x = \dfrac{\pi}{2} + k\pi$
then the asymptotes are $x = \dfrac{1}{2} + k$ for any integer k.

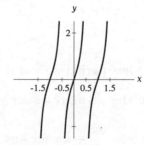

39. $y = -2\tan(x) + 1$ has period π, and the asymptotes are $x = \dfrac{\pi}{2} + k\pi$ where k is an integer.

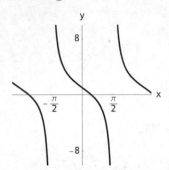

41. $y = -\tan(x - \pi/2)$ has period π, and if $x - \pi/2 = \dfrac{\pi}{2} + k\pi$ then the asymptotes are $x = \pi + k\pi$ or $x = k\pi$ for any integer k.

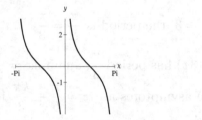

43. $y = \tan\left(\dfrac{\pi}{2}x - \dfrac{\pi}{2}\right)$ has period $\dfrac{\pi}{\pi/2}$ or 2, and if $\dfrac{\pi}{2}x - \dfrac{\pi}{2} = \dfrac{\pi}{2} + k\pi$ then the asymptotes are $x = 2 + 2k$ or $x = 2k$ for any integer k.

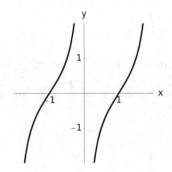

45. $y = \cot(x + \pi/4)$ has period π, and if $x + \dfrac{\pi}{4} = k\pi$ then the asymptotes are $x = -\dfrac{\pi}{4} + k\pi$ or $x = \dfrac{3\pi}{4} + k\pi$ for any integer k.

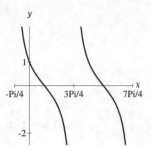

47. $y = \cot(x/2)$ has period 2π, and if $\dfrac{x}{2} = k\pi$ then the asymptotes are $x = 2k\pi$ for any integer k.

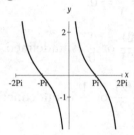

49. $y = -\cot(x + \pi/2)$ has period π, and if $x + \dfrac{\pi}{2} = k\pi$ then the asymptotes are $x = -\dfrac{\pi}{2} + k\pi$ or $x = \dfrac{\pi}{2} + k\pi$ for any integer k.

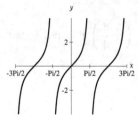

51. $y = \cot(2x - \pi/2) - 1$ has period $\dfrac{\pi}{2}$, and if $2x - \dfrac{\pi}{2} = k\pi$ then the asymptotes are $x = \dfrac{\pi}{4} + \dfrac{k\pi}{2}$ for any integer k.

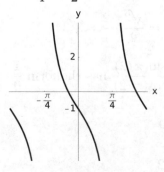

53. $y = 3\tan\left(x - \dfrac{\pi}{4}\right) + 2$

55. $y = -\cot\left(x + \dfrac{\pi}{2}\right) + 1$

57. Note, the period is $\dfrac{\pi}{2}$. So $\dfrac{\pi}{B} = \dfrac{\pi}{2}$ and $B = 2$.

The phase shift is $\dfrac{\pi}{4}$ and $A = 1$ since $\left(\dfrac{3\pi}{8}, 1\right)$ is a point on the graph.

An equation is $y = \tan\left(2\left(x - \dfrac{\pi}{4}\right)\right)$.

59. Note, the period is 2. So $\dfrac{\pi}{B} = 2$ and $B = \dfrac{\pi}{2}$.

Since the graph is reflected about the x-axis and $\left(\dfrac{1}{2}, -1\right)$ is a point on the graph, we find

$A = -1$. An equation is $y = -\tan\left(\dfrac{\pi}{2}x\right)$.

61. $f(g(-3)) = f(0) = \tan(0) = 0$

63. Undefined, since $\tan(\pi/2)$ is undefined and $g(h(f(\pi/2))) = g(h(\tan(\pi/2)))$

65. $f(g(h(x))) = f(g(2x)) = f(2x + 3) = \tan(2x + 3)$

67. $g(h(f(x))) = g(h(\tan(x))) = g(2\tan(x)) = 2\tan(x) + 3$

69. Note $m = \tan(\pi/4) = 1$. Since the line passes through $(2, 3)$, we get $y - 3 = 1 \cdot (x - 2)$. Solving for y, we obtain $y = x + 1$.

71. Note $m = \tan(\pi/3) = \sqrt{3}$. Since the line passes through $(3, -1)$, we get

$$y + 1 = \sqrt{3}(x - 3).$$

Solving for y, we obtain $y = \sqrt{3}x - 3\sqrt{3} - 1$.

73.

   a) Period is about 2.3 years

   b) It looks like the graph of a tangent function.

75. $\csc\alpha = \dfrac{1}{y}$, $\sec\alpha = \dfrac{1}{x}$, $\cot\alpha = \dfrac{x}{y}$

77. Period is $\dfrac{2\pi}{B} = \dfrac{2\pi}{2} = \pi$.

To find the asymptotes, solve

$$
\begin{aligned}
2x - \pi &= \frac{\pi}{2} + m\pi \\
2x &= \frac{\pi}{2} + (m + 1)\pi \\
x &= \frac{\pi}{4} + \frac{(m + 1)\pi}{2}
\end{aligned}
$$

where m is an integer. Let $k = m + 1$. Then the asymptotes are

$$x = \frac{\pi}{4} + \frac{k\pi}{2}$$

The range is $(-\infty, -3] \cup [3, \infty)$.

79. Let x be the distance from the building to the boss. Then

$$\tan 28° = \frac{432}{x}$$

and $x = 432/\tan 28° \approx 812$ feet.

81.

   a) First, use four tiles to make a 4-by-4 square. Then construct three more 4-by-4 square squares. Now, you have four 4-by-4 squares. Then put these four squares together to make a 8-by-8 square.

   b) By elimination, you will not be able to make a 6-by-6 square. There are only a few possibilities and none of them will make a 6-by-6 square.

For Thought

1. False, since the graph of $y = x + \sin x$ does not duplicate itself.

2. True, since the range of $y = \sin x$ is $[-1, 1]$ it follows that $y = x + \sin x$ oscillates about $y = x$.

3. True, since if x is small then the value of y in $y = \dfrac{1}{x}$ is a large number.

4. True, since $y = 0$ is the horizontal asymptote of $y = \dfrac{1}{x}$ and the x-axis is the graph of $y = 0$.

5. True, since the range of $y = \sin x$ is $[-1, 1]$ it follows that $y = \dfrac{1}{x} + \sin x$ oscillates about $y = \dfrac{1}{x}$.

6. True, since $\sin x = 1$ whenever $x = \dfrac{\pi}{2} + k2\pi$ and $\sin x = -1$ whenever $x = \dfrac{3\pi}{2} + k2\pi$, and since $\dfrac{1}{x}$ is approximately zero when x is a large numbers, then $\dfrac{1}{x} + \sin x = 0$ has many solutions in x for 0 is between 1 and -1.

7. False, since $\cos(\pi/6) + \cos(2 \cdot \pi/6) = \dfrac{\sqrt{3} + 1}{2} > 1$ then 1 is not the maximum of $y = \cos(x) + \cos(2x)$.

8. False, since $\sin(x) + \cos(x) = \sqrt{2}\sin\left(x + \dfrac{\pi}{4}\right)$ then the maximum value of $\sin(x) + \cos(x)$ is $\sqrt{2}$, and not 2.

9. True, since on the interval $[0, 2\pi]$ we find that $\sin(x) = 0$ for $x = 0, \pi, 2\pi$.

10. True, since $B = \pi$ and the period is $\dfrac{2\pi}{B} = \dfrac{2\pi}{\pi}$ or 2.

2.5 Exercises

1. For each x-coordinate, the y-coordinate of $y = x + \cos x$ is the sum of the y-coordinates of $y_1 = x$ and $y_2 = \cos x$. Note, the graph below oscillates about $y_1 = x$.

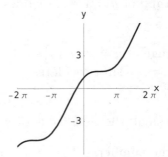

3. For each x-coordinate, the y-coordinate of $y = \dfrac{1}{x} - \sin x$ is obtained by subtracting the y-coordinate of $y_2 = \sin x$ from the y-coordinate of $y_1 = \dfrac{1}{x}$.

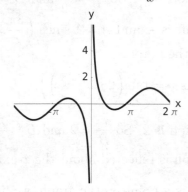

5. For each x-coordinate, the y-coordinate of $y = \dfrac{1}{2}x + \sin x$ is the sum of the y-coordinates of $y_1 = \dfrac{1}{2}x$ and $y_2 = \sin x$.

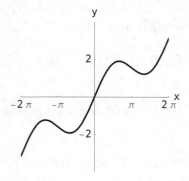

7. For each x-coordinate, the y-coordinate of $y = x^2 + \sin x$ is the sum of the y-coordinates of $y_1 = x^2$ and $y_2 = \sin x$. Note, $y_2 = \sin x$ oscillates about $y_1 = x^2$.

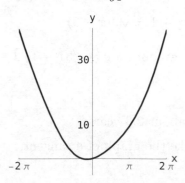

9. For each x-coordinate, the y-coordinate of $y = \sqrt{x} + \cos x$ is the sum of the y-coordinates of $y_1 = \sqrt{x}$ and $y_2 = \cos x$.

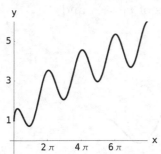

11. For each x-coordinate, the y-coordinate of $y = |x| + 2\sin x$ is the sum of the y-coordinates of $y_1 = |x|$ and $y_2 = 2\sin x$. Note, the graph below oscillates about $y_1 = |x|$.

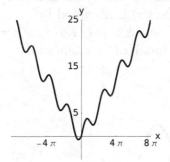

13. For each x-coordinate, the y-coordinate of $y = \cos(x) + 2\sin(x)$ is obtained by adding the y-coordinates of $y_1 = \cos x$ and $y_2 = 2\sin x$. Note, $y = \cos(x) + 2\sin(x)$ is a periodic function since it is the sum of two periodic functions.

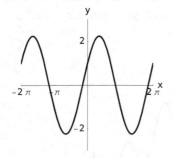

15. For each x-coordinate, the y-coordinate of $y = \sin(x) - \cos(x)$ is obtained by subtracting the y-coordinate of $y_2 = \cos x$ from the y-coordinate of $y_1 = \sin(x)$. Note, $y = \sin(x) - \cos(x)$ is a periodic function since it is the difference of two periodic functions.

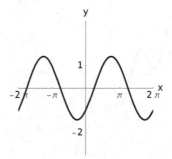

17. For each x-coordinate, the y-coordinate of $y = \sin(x) + \cos(2x)$ is obtained by adding the y-coordinates of $y_1 = \sin x$ and $y_2 = \cos(2x)$. Note, $y = \sin(x) + \cos(2x)$ is a periodic function since it is the sum of two periodic functions.

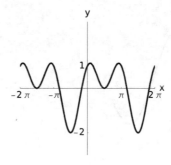

19. For each x-coordinate, the y-coordinate of $y = 2\sin(x) - \cos(2x)$ is obtained by subtracting the y-coordinate of $y_2 = \cos(2x)$ from the y-coordinate of $y_1 = 2\sin(x)$. Note, $y = 2\sin(x) - \cos(2x)$ is a periodic function with period 2π.

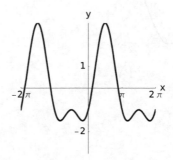

21. For each x-coordinate, the y-coordinate of $y = \sin(x) + \sin(2x)$ is obtained by adding the y-coordinates of $y_1 = \sin x$ and $y_2 = \sin(2x)$. Note, $y = \sin(x) + \sin(2x)$ is a periodic with period 2π.

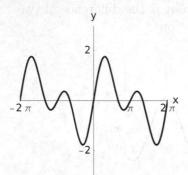

23. For each x-coordinate, the y-coordinate of $y = \sin(x) + \cos\left(\dfrac{x}{2}\right)$ is obtained by adding the y-coordinates of $y_1 = \sin x$ and $y_2 = \cos\left(\dfrac{x}{2}\right)$. Note, $y = \sin(x) + \cos\left(\dfrac{x}{2}\right)$ is a periodic function with period 4π.

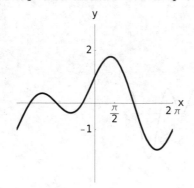

25. For each x-coordinate, the y-coordinate of $y = x + \cos(\pi x)$ is obtained by adding the y-coordinates of $y_1 = x$ and $y_2 = \cos(\pi x)$.

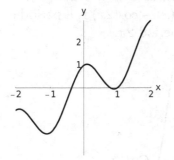

27. For each x-coordinate, the y-coordinate of $y = \dfrac{1}{x} + \cos(\pi x)$ is obtained by adding the y-coordinates of $y_1 = \dfrac{1}{x}$ and $y_2 = \cos(\pi x)$.

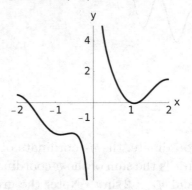

29. For each x-coordinate, the y-coordinate of $y = \sin(\pi x) - \cos(\pi x)$ is obtained by subtracting the y-coordinate of $y_2 = \cos(\pi x)$ from the y-coordinate of $y_1 = \sin(\pi x)$.

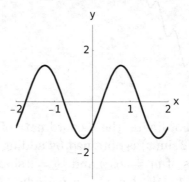

31. For each x-coordinate, the y-coordinate of $y = \sin(\pi x) + \sin(2\pi x)$ is obtained by adding the y-coordinates of $y_1 = \sin(\pi x)$ and $y_2 = \sin(2\pi x)$.

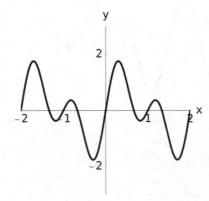

33. For each x-coordinate, the y-coordinate of $y = \cos\left(\dfrac{\pi}{2}x\right) + \sin(\pi x)$ is obtained by adding the y-coordinates of $y_1 = \cos\left(\dfrac{\pi}{2}x\right)$ and $y_2 = \sin(\pi x)$.

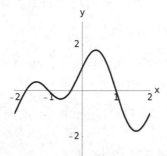

35. For each x-coordinate, the y-coordinate of $y = \sin(\pi x) + \sin\left(\dfrac{\pi}{2}x\right)$ is obtained by adding the y-coordinates of $y_1 = \sin(\pi x)$ and $y_2 = \sin\left(\dfrac{\pi}{2}x\right)$.

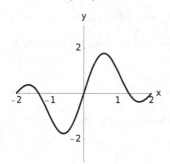

37. c, since the graph of of $y = x + \sin(6x)$ oscillates about the line $y = x$

39. a, since the graph of of $y = -x + \cos(0.5x)$ oscillates about the line $y = -x$

41. d, since the graph of of $y = \sin x + \cos(10x)$ is periodic and passes through $(0, 1)$

43. Since $x_\circ = -3$, $v_\circ = 4$, and $\omega = 1$, we obtain

$$x(t) \;=\; \dfrac{v_\circ}{\omega}\sin(\omega t) + x_\circ \cos(\omega t)$$
$$\;=\; 4\sin t - 3\cos t.$$

After $t = 3$ sec, the location of the weight is

$$x(3) = 4\sin 3 - 3\cos 3 \approx 3.5 \text{ cm}.$$

The period and amplitude of

$$x(t) = 4\sin t - 3\cos t$$

are $2\pi \approx 6.3$ sec and $\sqrt{4^2 + 3^2} = 5$ cm, respectively.

45. a) The graph of

$$P(x) = 1000(1.01)^x + 500\sin\left(\dfrac{\pi}{6}(x-4)\right) + 2000$$

for $1 \le x \le 60$ is given below

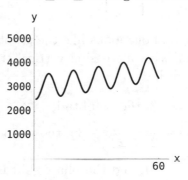

b) The graph for $1 \le x \le 600$ is given below

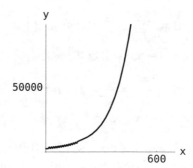

The graph looks like an exponential function.

47. By adding the ordinates of $y = x$ and $y = \sin(x)$, one can obtain the graph of $y = x + \sin(x)$ (which is given below).

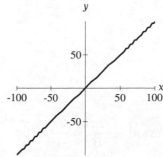

The graph above looks like the graph of $y = x$.

49. By adding the ordinates of $y = x$ and $y = \tan(x)$, we obtain the graph of $y = x + \tan(x)$.

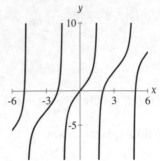

The graph above looks like the graph of $y = \tan x$ with increasing vertical translation on each cycle.

51. Since $A = -3$, the amplitude is 3.
Since $\dfrac{\pi x}{2} - \dfrac{\pi}{2} = \dfrac{\pi}{2}(x - 1)$, the period is $\dfrac{2\pi}{B} = \dfrac{2\pi}{\pi/2} = 4$, and the phase shift is 1.
The range is $[-3 + 7, 3 + 7]$ or $[4, 10]$

53. The period is $\dfrac{2\pi}{B} = \dfrac{2\pi}{\pi} = 2$, and the range is $(-\infty, -5] \cup [5, \infty)$.

55. The period is $\dfrac{\pi}{B} = \dfrac{\pi}{\pi} = 1$, and the range is $(-\infty, \infty)$.

57. Since there will be no 1's left after the 162nd house, the first house that cannot be numbered correctly is 163.

Review Exercises

1. 1

3. $\dfrac{\sin(\pi/4)}{\cos(\pi/4)} = \dfrac{\sqrt{2}/2}{\sqrt{2}/2} = 1$

5. $\dfrac{1}{\cos \pi} = \dfrac{1}{-1} = -1$

7. 0

9. Since $B = 1$, the period is $\dfrac{2\pi}{B} = \dfrac{2\pi}{1} = 2\pi$.

11. Since $B = 2$, the period is $\dfrac{\pi}{B} = \dfrac{\pi}{2}$.

13. Since $B = \pi$, the period is $\dfrac{2\pi}{B} = \dfrac{2\pi}{\pi} = 2$.

15. Since $B = \dfrac{1}{2}$, the period is $\dfrac{2\pi}{B} = \dfrac{2\pi}{1/2} = 4\pi$.

17. Domain $(-\infty, \infty)$, range $[-2, 2]$

19. To find the domain of $y = \tan(2x)$, solve $2x = \dfrac{\pi}{2} + k\pi$, and so the domain is $\left\{ x \mid x \neq \dfrac{\pi}{4} + \dfrac{k\pi}{2} \right\}$. The range is $(-\infty, \infty)$.

21. Since the zeros of $y = \cos(x)$ are $x = \dfrac{\pi}{2} + k\pi$, the domain of $y = \sec(x) - 2$ is $\left\{ x \mid x \neq \dfrac{\pi}{2} + k\pi \right\}$. The range is
$$(-\infty, -3] \cup [-1, \infty).$$

23. By using the zeros of $y = \sin(x)$ and by solving $\pi x = k\pi$, we get that the domain of $y = \cot(\pi x)$ is $\{ x \mid x \neq k \}$. The range of $y = \cot(\pi x)$ is $(-\infty, \infty)$.

25. By solving $2x = \dfrac{\pi}{2} + k\pi$, we get that the asymptotes of $y = \tan(2x)$ are
$$x = \dfrac{\pi}{4} + \dfrac{k\pi}{2}.$$

27. By solving $\pi x = k\pi$, we find that the asymptotes of $y = \cot(\pi x) + 1$ are $x = k$.

29. Solving $x - \dfrac{\pi}{2} = \dfrac{\pi}{2} + k\pi$, we get $x = \pi + k\pi$ or equivalently $x = k\pi$. Then the asymptotes of $y = \sec\left(x - \dfrac{\pi}{2} \right)$ are $x = k\pi$.

31. Solving $\pi x + \pi = k\pi$, we get $x = -1 + k$ or equivalently $x = k$. Then the asymptotes of $y = \csc(\pi x + \pi)$ are $x = k$.

33. Amplitude 3, since $B = 2$ the period is $\dfrac{2\pi}{B} = \dfrac{2\pi}{2}$ or equivalently π, phase shift is 0, and range is $[-3, 3]$. Five points are $(0, 0)$, $\left(\dfrac{\pi}{4}, 3\right)$, $\left(\dfrac{\pi}{2}, 0\right)$, $\left(\dfrac{3\pi}{4}, -3\right)$, and $(\pi, 0)$.

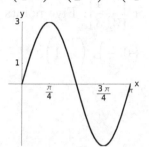

35. Amplitude 1, since $B = 1$ the period is $\dfrac{2\pi}{B} = \dfrac{2\pi}{1}$ or equivalently 2π, phase shift is $\dfrac{\pi}{6}$, and range is $[-1, 1]$. Five points are $\left(\dfrac{\pi}{6}, 1\right)$, $\left(\dfrac{2\pi}{3}, 0\right)$, $\left(\dfrac{7\pi}{6}, -1\right)$, $\left(\dfrac{5\pi}{3}, 0\right)$, and $\left(\dfrac{13\pi}{6}, 1\right)$.

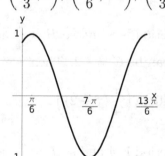

37. Amplitude 1, since $B = 1$ the period is $\dfrac{2\pi}{B} = \dfrac{2\pi}{1}$ or equivalently 2π, phase shift is $-\dfrac{\pi}{4}$, and range is $[-1 + 1, 1 + 1]$ or $[0, 2]$. Five points are $\left(-\dfrac{\pi}{4}, 2\right)$, $\left(\dfrac{\pi}{4}, 1\right)$, $\left(\dfrac{3\pi}{4}, 0\right)$, $\left(\dfrac{5\pi}{4}, 1\right)$, and $\left(\dfrac{7\pi}{4}, 2\right)$.

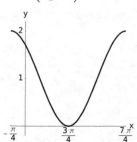

39. Amplitude 1, since $B = \dfrac{\pi}{2}$ the period is $\dfrac{2\pi}{B} = \dfrac{2\pi}{\pi/2}$ or equivalently 4, phase shift is 0, and range is $[-1, 1]$. Five points are $(0, 1)$, $(1, 0)$, $(2, -1)$, $(3, 0)$, and $(4, 1)$.

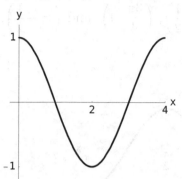

41. Note, $y = \sin\left(2\left(x + \dfrac{\pi}{2}\right)\right)$. Then amplitude is 1, since $B = 2$ the period is $\dfrac{2\pi}{B} = \dfrac{2\pi}{2}$ or equivalently π, phase shift is $-\dfrac{\pi}{2}$, and range is $[-1, 1]$. Five points are $\left(-\dfrac{\pi}{2}, 0\right)$, $\left(-\dfrac{\pi}{4}, 1\right)$, $(0, 0)$, $\left(\dfrac{\pi}{4}, -1\right)$, and $\left(\dfrac{\pi}{2}, 0\right)$.

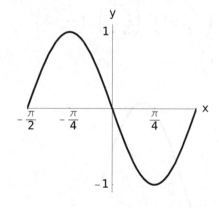

43. Amplitude $\frac{1}{2}$, since $B = 2$ the period is $\frac{2\pi}{B} = \frac{2\pi}{2}$ or equivalently π, phase shift is 0, and range is $\left[-\frac{1}{2}, \frac{1}{2}\right]$. Five points are $\left(0, -\frac{1}{2}\right)$, $\left(\frac{\pi}{4}, 0\right)$, $\left(\frac{\pi}{2}, \frac{1}{2}\right)$, $\left(\frac{3\pi}{4}, 0\right)$, and $\left(\pi, -\frac{1}{2}\right)$.

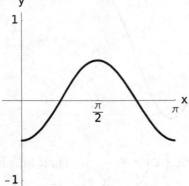

45. Note, $y = -2\sin\left(2\left(x + \frac{\pi}{6}\right)\right) + 1$.

Then amplitude is 2, since $B = 2$ the period is $\frac{2\pi}{B} = \frac{2\pi}{2}$ or π, phase shift is $-\frac{\pi}{6}$, and range is $[-2 + 1, 2 + 1]$ or $[-1, 3]$. Five points are $\left(-\frac{\pi}{6}, 1\right)$, $\left(\frac{\pi}{12}, -1\right)$, $\left(\frac{\pi}{3}, 1\right)$, $\left(\frac{7\pi}{12}, 3\right)$, and $\left(\frac{5\pi}{6}, 1\right)$.

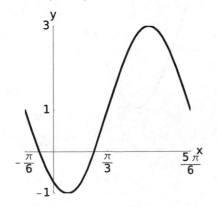

47. Note, $y = \frac{1}{3}\cos\left(\pi\left(x + 1\right)\right) + 1$.

Then amplitude is $\frac{1}{3}$, since $B = \pi$ the period is $\frac{2\pi}{B} = \frac{2\pi}{\pi}$ or 2, phase shift is -1, and range is $\left[-\frac{1}{3} + 1, \frac{1}{3} + 1\right]$ or $\left[\frac{2}{3}, \frac{4}{3}\right]$. Five points are $\left(-1, \frac{4}{3}\right)$, $\left(-\frac{1}{2}, 1\right)$, $\left(0, \frac{2}{3}\right)$, $\left(\frac{1}{2}, 1\right)$, and $\left(1, \frac{4}{3}\right)$.

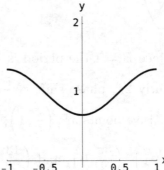

49. Note, $A = 10$, period is 8 and so $B = \frac{\pi}{4}$, and the phase shift can be $C = -2$. Then
$$y = 10\sin\left(\frac{\pi}{4}\left(x + 2\right)\right).$$

51. Note, $A = 20$, period is 4 and so $B = \frac{\pi}{2}$, and the phase shift can be $C = 1$, and vertical shift is 10 units up or $D = 10$. Then
$$y = 20\sin\left(\frac{\pi}{2}\left(x - 1\right)\right) + 10.$$

53. Since $B = 3$, the period is $\dfrac{\pi}{B} = \dfrac{\pi}{3}$. To find the asymptotes, let $3x = \dfrac{\pi}{2} + k\pi$. Then the asymptotes are $x = \dfrac{\pi}{6} + \dfrac{k\pi}{3}$. The range is $(-\infty, \infty)$.

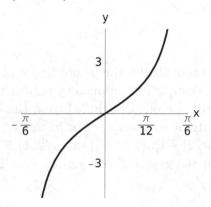

55. Since $B = 2$, the period is $\dfrac{\pi}{B} = \dfrac{\pi}{2}$. To find the asymptotes, let $2x + \pi = \dfrac{\pi}{2} + k\pi$. Then $x = -\dfrac{\pi}{4} + \dfrac{k\pi}{2}$ or equivalently we get $x = \dfrac{\pi}{4} + \dfrac{k\pi}{2}$, which are the asymptotes. The range is $(-\infty, \infty)$.

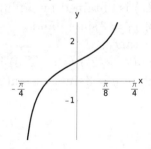

57. Since $B = \dfrac{1}{2}$, the period is $\dfrac{2\pi}{B} = \dfrac{2\pi}{1/2}$ or 4π. To find the asymptotes, let $\dfrac{1}{2}x = \dfrac{\pi}{2} + k\pi$. Then $x = \pi + 2k\pi$ which are the asymptotes. The range is $(-\infty, -1] \cup [1, \infty)$.

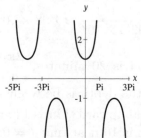

59. Since $B = 2$, the period is $\dfrac{\pi}{B} = \dfrac{\pi}{2}$. To find the asymptotes, let $2x = k\pi$. Then the asymptotes are $x = \dfrac{k\pi}{2}$. The range is $(-\infty, \infty)$.

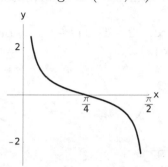

61. Since $B = 2$, the period is $\dfrac{\pi}{B} = \dfrac{\pi}{2}$. To find the asymptotes, let $2x + \dfrac{\pi}{3} = k\pi$. Then $x = -\dfrac{\pi}{6} + \dfrac{k\pi}{2}$ or equivalently $x = \dfrac{\pi}{3} + \dfrac{k\pi}{2}$, which are the asymptotes. The range is $(-\infty, \infty)$.

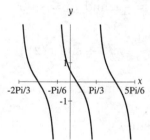

63. Since $B = 2$, the period is $\dfrac{2\pi}{B} = \dfrac{2\pi}{2}$ or π. To obtain the asymptotes, let $2x + \pi = k\pi$ or equivalently $2x = k\pi$. Then the asymptotes are $x = \dfrac{k\pi}{2}$. The range is $(-\infty, -1/3] \cup [1/3, \infty)$.

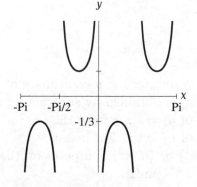

65. Since $B = 1$, the period is $\dfrac{2\pi}{B} = \dfrac{2\pi}{1}$ or 2π.

To find the asymptotes, let $x + \dfrac{\pi}{4} = \dfrac{\pi}{2} + k\pi$.

Then $x = \dfrac{\pi}{4} + k\pi$ which are the asymptotes.

The range is $(-\infty, -2 - 1] \cup [2 - 1, \infty)$ or $(-\infty, -3] \cup [1, \infty)$.

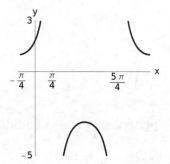

67. For each x-coordinate, the y-coordinate of $y = \dfrac{1}{2}x + \sin(x)$ is obtained by adding the y-coordinates of $y_1 = \dfrac{1}{2}x$ and $y_2 = \sin x$.

For instance, $\left(\pi/2, \dfrac{1}{2} \cdot \pi/2 + \sin(\pi/2)\right)$

or $(\pi/2, \pi/4 + 1)$ is a point on the

graph of $y = \dfrac{1}{2}x + \sin(x)$.

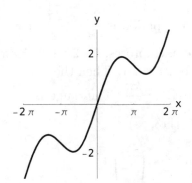

69. For each x-coordinate, the y-coordinate of $y = x^2 - \sin(x)$ is obtained by subtracting the y-coordinate of $y_2 = \sin x$ from the y-coordinate of $y_1 = x^2$. For instance, $(\pi, \pi^2 - \sin(\pi))$ or (π, π^2) is a point on the graph of $y = x^2 - \sin(x)$.

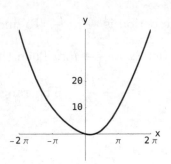

71. For each x-coordinate, the y-coordinate of $y = \sin(x) - \sin(2x)$ is obtained by subtracting the y-coordinate of $y_2 = \sin(2x)$ from the y-coordinate of $y_1 = \sin(x)$. For instance, $(\pi/2, \sin(\pi/2) - \sin(2 \cdot \pi/2))$ or $(\pi/2, 1)$ is a point on the graph of $y = \sin(x) - \sin(2x)$.

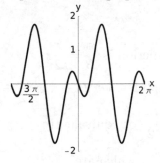

73. For each x-coordinate, the y-coordinate of $y = 3\sin(x) + \sin(2x)$ is obtained by adding the y-coordinates of $y_1 = 3\sin(x)$ and $y_2 = \sin(2x)$. For instance, $(\pi/2, 3\sin(\pi/2) + \sin(2 \cdot \pi/2))$ or $(\pi/2, 3)$ is a point on the graph of $y = 3\sin(x) + \sin(2x)$.

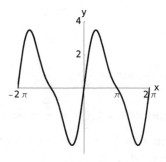

75. The period is $\dfrac{1}{92.3 \times 10^6} \approx 1.08 \times 10^{-8}$ sec

77. Since the period is 20 minutes, $\dfrac{2\pi}{b} = 20$ or $b = \dfrac{\pi}{10}$. Since the depth is between 12 ft and 16 ft, the vertical upward shift is 14 and $a = 2$. Since the depth is 16 ft at time $t = 0$, one can assume there is a left shift of 5 minutes. An equation is

$$y = 2\sin\left(\frac{\pi}{10}(x+5)\right) + 14$$

and its graph is given on the next page.

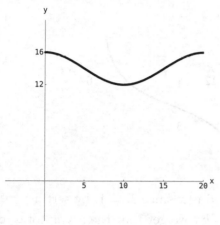

79. $9, $99, $999, $9,999, $99,999, $999,999, $9,999,999 $99,999,999, $999,999,999

Chapter 2 Test

1. Period $2\pi/3$, range $[-1, 1]$, amplitude 1, some points are $(0, 0)$, $\left(\frac{\pi}{6}, 1\right)$,

$\left(\frac{\pi}{3}, 0\right)$, $\left(\frac{\pi}{2}, -1\right)$, $\left(\frac{2\pi}{3}, 0\right)$

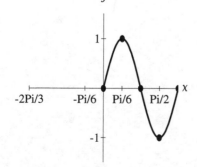

2. Period 2π, range $[-1, 1]$, amplitude 1, some points are $\left(-\frac{\pi}{4}, 1\right)$, $\left(\frac{\pi}{4}, 0\right)$, $\left(\frac{3\pi}{4}, -1\right)$, $\left(\frac{5\pi}{4}, 0\right)$, $\left(\frac{7\pi}{4}, 1\right)$

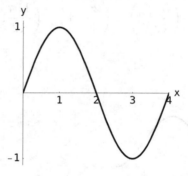

3. Since $B = \frac{\pi}{2}$, the period is 4.

The range is $[-1, 1]$ and amplitude is 1. Some points are $(0, 0)$, $(1, 1)$, $(2, 0)$, $(3, -1)$, $(4, 0)$

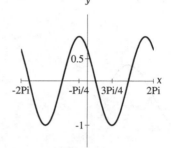

4. Period 2π, range $[-2, 2]$, amplitude 2, some points are $(\pi, -2)$, $\left(\frac{3\pi}{2}, 0\right)$, $(2\pi, 2)$, $\left(\frac{5\pi}{2}, 0\right)$, $(3\pi, -2)$

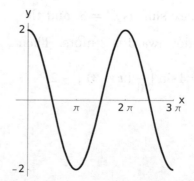

5. Period 2π, range $\left[-\dfrac{1}{2}, \dfrac{1}{2}\right]$, amplitude $\dfrac{1}{2}$, some

points are $\left(\dfrac{\pi}{2}, 0\right)$, $\left(\pi, -\dfrac{1}{2}\right)$, $\left(\dfrac{3\pi}{2}, 0\right)$,

$\left(2\pi, \dfrac{1}{2}\right)$, $\left(\dfrac{5\pi}{2}, 0\right)$

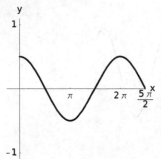

6. Period π since $B = 2$, range is $[-1 + 3, 1 + 3]$

or $[2, 4]$, amplitude is 1, some points

are $\left(\dfrac{\pi}{6}, 2\right)$, $\left(\dfrac{5\pi}{12}, 3\right)$, $\left(\dfrac{2\pi}{3}, 4\right)$, $\left(\dfrac{11\pi}{12}, 3\right)$,

$\left(\dfrac{7\pi}{6}, 2\right)$

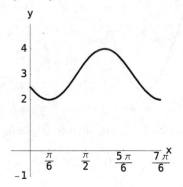

7. Note, amplitude is $A = 4$, period is 12 and

so $B = \dfrac{\pi}{6}$, phase shift is $C = 3$, and the

vertical shift downward is 2 units. Then

$$y = 4\sin\left(\dfrac{\pi}{6}(x - 3)\right) - 2.$$

8. The period is $\dfrac{\pi}{3}$ since $B = 3$, by setting

$3x = \dfrac{\pi}{2} + k\pi$ we get that the asymptotes are

$x = \dfrac{\pi}{6} + \dfrac{k\pi}{3}$, and the range is $(-\infty, \infty)$.

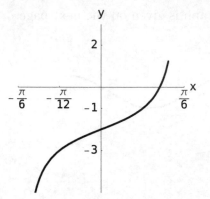

9. The period is π since $B = 1$, by setting

$x + \dfrac{\pi}{2} = k\pi$ we get that the asymptotes are

$x = -\dfrac{\pi}{2} + k\pi$, $x = -\dfrac{\pi}{2} + \pi + k\pi$, or $x = \dfrac{\pi}{2} + k\pi$,

and the range is $(-\infty, \infty)$.

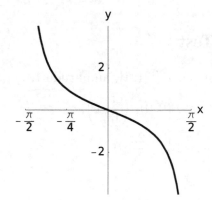

10. The period is $\dfrac{2\pi}{1}$ or 2π, by setting $x - \pi = \dfrac{\pi}{2} + k\pi$ we get $x = \dfrac{3\pi}{2} + k\pi$ and equivalently the asymptotes are $x = \dfrac{\pi}{2} + k\pi$, and the range is $(-\infty, -2] \cup [2, \infty)$.

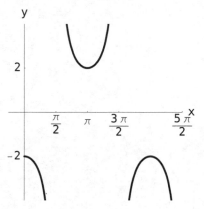

11. The period is $\dfrac{2\pi}{1}$ or 2π, by setting $x - \dfrac{\pi}{2} = k\pi$ we get that the asymptotes are $x = \dfrac{\pi}{2} + k\pi$, and the range is $(-\infty, -1+1] \cup [1+1, \infty)$ or $(-\infty, 0] \cup [2, \infty)$.

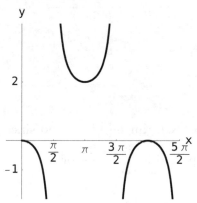

12. For each x-coordinate, the y-coordinate of $y = 3\sin(x) + \cos(2x)$ is obtained by adding the y-coordinates of $y_1 = 3\sin(x)$ and $y_2 = \cos(2x)$. For instance, $(\pi/2, 3\sin(\pi/2) + \cos(2 \cdot \pi/2)) = (\pi/2, 3 + (-1))$ or $(\pi/2, 2)$ is a point on the graph of $y = 3\sin(x) + \cos(2x)$.

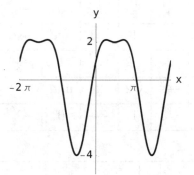

13. Since the pH oscillates between 7.2 and 7.8, there is a vertical upward shift of 7.5 and $a = 0.3$. Note, the period is 4 days. Thus, $\dfrac{2\pi}{b} = 4$ or $b = \dfrac{\pi}{2}$. Since the pH is 7.2 on day 13, the pH is 7.5 on day 14. We can assume a right shift of 14 days. Hence, an equation is

$$y = 0.3\sin\left(\dfrac{\pi}{2}(x - 14)\right) + 7.5.$$

A graph of one cycle is given below.

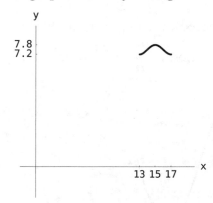

Tying It All Together Chapters P-2

1.

θ deg	0	30	45	60	90	120	135	150	180
θ rad	0	$\frac{\pi}{6}$	$\frac{\pi}{4}$	$\frac{\pi}{3}$	$\frac{\pi}{2}$	$\frac{2\pi}{3}$	$\frac{3\pi}{4}$	$\frac{5\pi}{6}$	π
$\sin\theta$	0	$\frac{1}{2}$	$\frac{\sqrt{2}}{2}$	$\frac{\sqrt{3}}{2}$	1	$\frac{\sqrt{3}}{2}$	$\frac{\sqrt{2}}{2}$	$\frac{1}{2}$	0
$\cos\theta$	1	$\frac{\sqrt{3}}{2}$	$\frac{\sqrt{2}}{2}$	$\frac{1}{2}$	0	$-\frac{1}{2}$	$-\frac{\sqrt{2}}{2}$	$-\frac{\sqrt{3}}{2}$	-1
$\tan\theta$	0	$\frac{\sqrt{3}}{3}$	1	$\sqrt{3}$	und	$-\sqrt{3}$	-1	$-\frac{\sqrt{3}}{3}$	0
$\csc\theta$	und	2	$\sqrt{2}$	$\frac{2\sqrt{3}}{3}$	1	$\frac{2\sqrt{3}}{3}$	$\sqrt{2}$	2	und
$\sec\theta$	1	$\frac{2\sqrt{3}}{3}$	$\sqrt{2}$	2	und	-2	$-\sqrt{2}$	$-\frac{2\sqrt{3}}{3}$	-1
$\cot\theta$	und	$\sqrt{3}$	1	$\frac{\sqrt{3}}{3}$	0	$-\frac{\sqrt{3}}{3}$	-1	$-\sqrt{3}$	und

2.

θ rad	π	$\frac{7\pi}{6}$	$\frac{5\pi}{4}$	$\frac{4\pi}{3}$	$\frac{3\pi}{2}$	$\frac{5\pi}{3}$	$\frac{7\pi}{4}$	$\frac{11\pi}{6}$	2π
θ deg	180	210	225	240	270	300	315	330	360
$\sin\theta$	0	$-\frac{1}{2}$	$-\frac{\sqrt{2}}{2}$	$-\frac{\sqrt{3}}{2}$	-1	$-\frac{\sqrt{3}}{2}$	$-\frac{\sqrt{2}}{2}$	$-\frac{1}{2}$	0
$\cos\theta$	-1	$-\frac{\sqrt{3}}{2}$	$-\frac{\sqrt{2}}{2}$	$-\frac{1}{2}$	0	$\frac{1}{2}$	$\frac{\sqrt{2}}{2}$	$\frac{\sqrt{3}}{2}$	1
$\tan\theta$	0	$\frac{\sqrt{3}}{3}$	1	$\sqrt{3}$	und	$-\sqrt{3}$	-1	$-\frac{\sqrt{3}}{3}$	0
$\csc\theta$	und	-2	$-\sqrt{2}$	$-\frac{2\sqrt{3}}{3}$	-1	$-\frac{2\sqrt{3}}{3}$	$-\sqrt{2}$	-2	und
$\sec\theta$	-1	$-\frac{2\sqrt{3}}{3}$	$-\sqrt{2}$	-2	und	2	$\sqrt{2}$	$\frac{2\sqrt{3}}{3}$	1
$\cot\theta$	und	$\sqrt{3}$	1	$\frac{\sqrt{3}}{3}$	0	$-\frac{\sqrt{3}}{3}$	-1	$-\sqrt{3}$	und

3. Domain $(-\infty, \infty)$, range $[-1, 1]$, and since

$B = 2$ the period is $\dfrac{2\pi}{B} = \dfrac{2\pi}{2}$ or π.

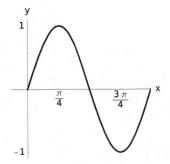

4. Domain $(-\infty, \infty)$, range $[-1, 1]$, and since

$B = 2$ the period is $\dfrac{2\pi}{B} = \dfrac{2\pi}{2}$ or π.

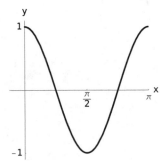

5. By setting $2x \neq \dfrac{\pi}{2} + k\pi$, the domain is $\left\{ x \ : \ x \neq \dfrac{\pi}{4} + \dfrac{k\pi}{2} \right\}$, range is $(-\infty, \infty)$, and the period is $\dfrac{\pi}{B} = \dfrac{\pi}{2}$.

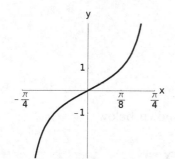

6. By setting $2x \neq \dfrac{\pi}{2} + k\pi$, we find that the domain is $\left\{ x \ : \ x \neq \dfrac{\pi}{4} + \dfrac{k\pi}{2} \right\}$, the range is $(-\infty, -1] \cup [1, \infty)$, and the period is $\dfrac{2\pi}{B} = \dfrac{2\pi}{2}$ or π.

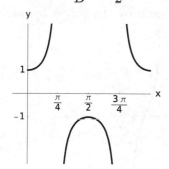

7. By setting $2x \neq k\pi$, we find that the domain is $\left\{ x \ : \ x \neq \dfrac{k\pi}{2} \right\}$, the range is $(-\infty, -1] \cup [1, \infty)$, and the period is $\dfrac{2\pi}{B} = \dfrac{2\pi}{2}$ or π.

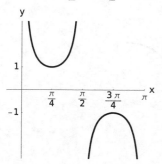

8. Domain $(-\infty, \infty)$, range $[-1, 1]$, and since $B = \pi$ the period is $\dfrac{2\pi}{B} = \dfrac{2\pi}{\pi}$ or 2.

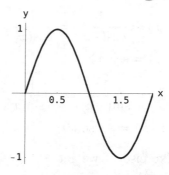

9. Domain $(-\infty, \infty)$, range $[-1, 1]$, and since $B = \pi$ the period is $\dfrac{2\pi}{B} = \dfrac{2\pi}{\pi}$ or 2.

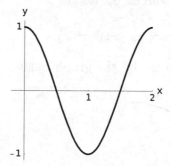

10. By setting $\pi x \neq \dfrac{\pi}{2} + k\pi$, the domain is $\left\{ x \ : \ x \neq \dfrac{1}{2} + k \right\}$, range is $(-\infty, \infty)$, and the period is $\dfrac{\pi}{B} = \dfrac{\pi}{\pi}$ or 1.

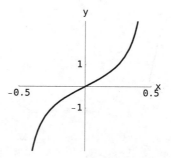

11. Odd, since $\sin(-x) = -\sin(x)$

12. Even, since $\cos(-x) = \cos(x)$

13. Odd, since $\tan(-x) = \dfrac{\sin(-x)}{\cos(-x)} = \dfrac{-\sin(x)}{\cos(x)} = -\tan(x)$, i.e., $\tan(-x) = -\tan(x)$

14. Odd, since $\cot(-x) = \dfrac{\cos(-x)}{\sin(-x)} = \dfrac{\cos(x)}{-\sin(x)} =$
$-\cot(x)$, i.e., $\cot(-x) = -\cot(x)$

15. Even, since $\sec(-x) = \dfrac{1}{\cos(-x)} = \dfrac{1}{\cos(x)} =$
$\sec(x)$, i.e., $\sec(-x) = \sec(x)$

16. Odd, since $\csc(-x) = \dfrac{1}{\sin(-x)} = \dfrac{1}{-\sin(x)} =$
$-\csc(x)$, i.e., $\csc(-x) = -\csc(x)$

17. Even, since by Exercise 11 we get
$\sin^2(-x) = (\sin(-x))^2 =$
$(-\sin(x))^2 = (\sin(x))^2$,
i.e., $\sin^2(-x) = \sin^2(x)$

18. Even, since by Exercise 12 we get
$\cos^2(-x) = (\cos(-x))^2 =$
$(\cos(x))^2$, i.e., $\cos^2(-x) = \cos^2(x)$

19. Increasing, as shown by the graph below.

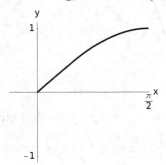

20. Decreasing, as shown by the graph below.

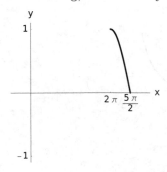

21. Increasing, as shown below.

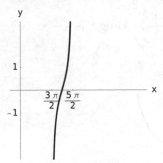

22. Decreasing, as shown below.

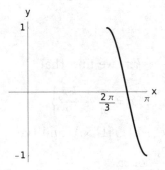

23. Decreasing, as shown by the graph below.

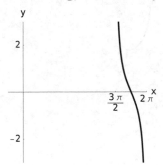

24. Increasing, as shown by the graph below.

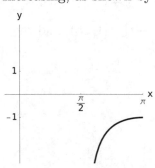

For Thought

1. False, for example $\sin 0 = 0$ but $\cos 0 = 1$.

2. True, since $\dfrac{\sin(x)}{\cos(x)} \cdot \dfrac{\cos(x)}{\sin(x)} = 1$.

3. True, since $\dfrac{2\sin(x)\cos(x)}{\sin(x)} = 2\cos(x)$.

4. False, since $(\sin x + \cos x)^2 = 1 + 2\sin(x)\cos(x) \neq 1 = \sin^2 x + \cos^2 x$.

5. False, since $\tan 1 = \sqrt{\sec^2(1) - 1}$.

6. False, since $\sin^2(-6) = (-\sin(6))^2 = \sin^2(6)$.

7. False, since $\cos(-9) = \cos(9)$ implies $\cos^3(-9) = \cos^3(9)$.

8. True, since $\sin(x)\csc(x) = 1$.

9. False, since $\sin^2(x) + \cos^2(x) = 1$.

10. True, since $(1-\sin x)(1+\sin x) = (1-\sin^2 x)$.

3.1 Exercises

1. Pythagorean

3. odd

5. odd

7. reciprocal

9. $\dfrac{1/\cos(x)}{\sin(x)/\cos(x)} = \dfrac{1}{\cos(x)} \cdot \dfrac{\cos(x)}{\sin(x)} =$
$\dfrac{1}{\sin(x)} = \csc(x)$

11.
$$\dfrac{\sin(x)}{1/\sin(x)} + \dfrac{\cos(x)}{1/\cos(x)} = \sin^2(x) + \cos^2(x) = 1$$

13. $1 - \sin^2 \alpha = \cos^2 \alpha$

15. $(\sin\beta + 1)(\sin\beta - 1) = \sin^2\beta - 1 = -\cos^2\beta$

17.
$$\dfrac{1 + \cos\alpha \cdot \dfrac{\sin\alpha}{\cos\alpha} \cdot \dfrac{1}{\sin\alpha}}{1/\sin\alpha} = \dfrac{1+1}{1/\sin\alpha} = 2\sin\alpha$$

19. Since $\cot^2 x = \csc^2 x - 1$, we get
$$\cot x = \pm\sqrt{\csc^2 x - 1}.$$

21. $\sin(x) = \dfrac{1}{\csc(x)} = \dfrac{1}{\pm\sqrt{1 + \cot^2(x)}}$

23. Since $\cot^2(x) = \csc^2(x) - 1$, we obtain
$$\tan(x) = \dfrac{1}{\cot(x)} = \dfrac{1}{\pm\sqrt{\csc^2(x) - 1}}$$

25. Since $\sec\alpha = \sqrt{1 + \left(\dfrac{1}{2}\right)^2} = \dfrac{\sqrt{5}}{2}$, we get
$$\cos\alpha = \dfrac{2}{\sqrt{5}} = \dfrac{2\sqrt{5}}{5}, \quad \sin\alpha = \sqrt{1 - \left(\dfrac{2}{\sqrt{5}}\right)^2} =$$
$\dfrac{1}{\sqrt{5}}$ or $\sin\alpha = \dfrac{\sqrt{5}}{5}$, $\csc\alpha = \sqrt{5}$, and $\cot\alpha = 2$.

27. Since $\sin\alpha = -\sqrt{1 - \left(-\dfrac{\sqrt{3}}{5}\right)^2} =$
$-\sqrt{1 - \dfrac{3}{25}} = -\dfrac{\sqrt{22}}{5}$, we find
$$\csc\alpha = -\dfrac{5}{\sqrt{22}} \text{ or } \csc\alpha = -\dfrac{5\sqrt{22}}{22},$$
$$\sec\alpha = -\dfrac{5}{\sqrt{3}} = -\dfrac{5\sqrt{3}}{3},$$
$$\tan\alpha = \dfrac{-\sqrt{22}/5}{-\sqrt{3}/5} = \dfrac{\sqrt{22}}{\sqrt{3}} = \dfrac{\sqrt{66}}{3}, \text{ and}$$
$$\cot\alpha = \dfrac{\sqrt{3}}{\sqrt{22}} = \dfrac{\sqrt{66}}{22}.$$

29. Since α is in Quadrant IV, we get
$$\csc\alpha = -\sqrt{1 + \left(-\dfrac{1}{3}\right)^2} = -\dfrac{\sqrt{10}}{3},$$
$$\sin\alpha = -\dfrac{3}{\sqrt{10}} = -\dfrac{3\sqrt{10}}{10},$$
$$\cos\alpha = \sqrt{1 - \left(-\dfrac{3}{\sqrt{10}}\right)^2} =$$
$$\sqrt{1 - \dfrac{9}{10}} = \dfrac{1}{\sqrt{10}} = \dfrac{\sqrt{10}}{10},$$
$\sec\alpha = \sqrt{10}$, and $\tan\alpha = -3$.

31. $(-\sin x) \cdot (-\cot x) = \sin(x) \cdot \dfrac{\cos x}{\sin x} = \cos(x)$

33. $\sin(y) + (-\sin(y)) = 0$

35. $\dfrac{\sin(x)}{\cos(x)} + \dfrac{-\sin(x)}{\cos(x)} = 0$

37. $(1 + \sin \alpha)(1 - \sin \alpha) = 1 - \sin^2 \alpha = \cos^2 \alpha$

39. $(-\sin \beta)(\cos \beta)(1/\sin \beta) = -\cos \beta$

41. Odd, since $\sin(-y) = -\sin(y)$ for any y, even if $y = 2x$.

43. Neither, since $f(-\pi/6) \neq f(\pi/6)$ and $f(-\pi/6) \neq -f(\pi/6)$.

45. Even, since $\sec^2(-t) - 1 = \sec^2(t) - 1$.

47. Even, $f(-\alpha) = 1 + \sec(-\alpha) = 1 + \sec(\alpha) = f(\alpha)$

49. Even, $f(-x) = \dfrac{\sin(-x)}{-x} = \dfrac{-\sin(x)}{-x} = f(x)$

51. Odd, $f(-x) = -x + \sin(-x) = -x - \sin(x) = -f(x)$

53. If $\gamma = \pi/3$, then $(\sin(\pi/3) + \cos(\pi/3))^2 = \left(\dfrac{\sqrt{3}}{2} + \dfrac{1}{2}\right)^2 = \dfrac{(\sqrt{3}+1)^2}{4} = \dfrac{4 + 2\sqrt{3}}{4} \neq 1$ while $\sin^2(\pi/3) + \cos^2(\pi/3) = 1$. Thus, it is not an identity.

55. If $\beta = \pi/6$, then

$$(1 + \sin(\pi/6))^2 = \left(1 + \frac{1}{2}\right)^2 = \left(\frac{3}{2}\right)^2 = \frac{9}{4}$$

and

$$1 + \sin^2(\pi/6) = 1 + \left(\frac{1}{2}\right)^2 = \frac{5}{4}.$$

Thus, it is not an identity.

57. If $\alpha = 7\pi/6$, then $\sin(7\pi/6) = -\dfrac{1}{2}$ while $\sqrt{1 - \cos^2\left(\dfrac{7\pi}{6}\right)}$ is a positive number. Thus, it is not an identity.

59. If $y = \pi/6$, then $\sin(\pi/6) = \dfrac{1}{2}$ while $\sin(-\pi/6) = -\dfrac{1}{2}$. Thus, it is not an identity.

61. If $y = \pi/6$, then $\cos^2(\pi/6) - \sin^2(\pi/6) = \left(\dfrac{\sqrt{3}}{2}\right)^2 - \left(\dfrac{1}{2}\right)^2 = \dfrac{3}{4} - \dfrac{1}{4} = \dfrac{1}{2}$ and

$$\sin(2 \cdot \pi/6) = \sin(\pi/3) = \frac{\sqrt{3}}{2}.$$

Thus, it is not an identity.

63. $1 - \dfrac{1}{\cos^2(x)} = 1 - \sec^2(x) = -\tan^2(x)$

$-\sin^2(x)\cos^3(x)$

65. $\dfrac{-(\tan^2 t + 1)}{\sec^2 t} = \dfrac{-\sec^2 t}{\sec^2 t} = -1$

67. $\dfrac{(1 - \cos^2 \alpha) - \cos^2 \alpha}{1 - 2\cos^2 \alpha} = \dfrac{1 - 2\cos^2 \alpha}{1 - 2\cos^2 \alpha} = 1$

69. $\dfrac{\tan x(\tan^2 x - \sec^2 x)}{-\cot x} = \dfrac{\tan x(-1)}{-\cot x} = \tan^2 x$

71.

$$\frac{1}{\sin^3 x} - \frac{\cos^2(x)/\sin^2(x)}{\sin x} = \frac{1}{\sin^3 x} - \frac{\cos^2(x)}{\sin^3 x} =$$

$$\frac{1 - \cos^2 x}{\sin^3 x} = \frac{\sin^2 x}{\sin^3 x} = \frac{1}{\sin x} = \csc x$$

73. $(\sin^2 x - \cos^2 x)(\sin^2 x + \cos^2 x) = (\sin^2 x - \cos^2 x)(1) = \sin^2 x - \cos^2 x$

75. Let b be the hypotenuse of the smaller right triangle shown below. Let a be the side opposite the head angle θ. Then $a + b = r$, the radius of the wheel.

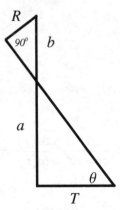

The two right triangles are similar triangles. Applying right triangle trigonometry,

$$b = \frac{R}{\cos \theta}.$$

Consequently,

$$a = r - b = r - \frac{R}{\cos \theta}.$$

Thus,

$$T = \frac{a}{\tan \theta} = \frac{r - \frac{R}{\cos \theta}}{\tan \theta}.$$

Simplifying, we obtain $T = \dfrac{r\cos\theta - R}{\sin \theta}$.

79. $360° - 35° = 325°$

81. $A = \theta r^2/2 = 15° \cdot \dfrac{\pi}{180°} \cdot \dfrac{6^2}{2} = \dfrac{3\pi}{2}$ ft^2

83. a) 0   **b)** $\dfrac{\sqrt{2}}{2}$ **c)** Undefined

　　 d) 0   **e)** $\dfrac{2\sqrt{3}}{3}$ **f)** 1

　　 g) $\dfrac{\sqrt{3}}{3}$ **h)** -1

85. Let x be the number of ants. Then

$$x = 10a + 6 = 7b + 2 = 11c + 2 = 13d + 2$$

for some positive integers a, b, c, d.

The smallest positive x satisfying the system of equations is $x = 4006$ ants.

For Thought

1. True, $\dfrac{\sin x}{1/\sin x} = \sin x \cdot \dfrac{\sin x}{1} = \sin^2 x$.

2. False, if $x = \pi/3$ then $\dfrac{\cot(\pi/3)}{\tan(\pi/3)} =$

$\dfrac{\sqrt{3}/3}{\sqrt{3}} = \dfrac{1}{3}$ and $\tan^2(\pi/3) = (\sqrt{3})^2 = 3$.

3. True, $\dfrac{1/\cos x}{1/\sin x} = \dfrac{1}{\cos x} \cdot \dfrac{\sin x}{1} = \dfrac{\sin x}{\cos x} = \tan x$.

4. True, $\sin x \cdot \dfrac{1}{\cos x} = \dfrac{\sin x}{\cos x} = \tan x$.

5. True, $\dfrac{\cos x}{\cos x} + \dfrac{\sin x}{\cos x} = 1 + \tan x$.

6. False, if $x = \pi/4$ then

$$\sec(\pi/4) + \dfrac{\sin(\pi/4)}{\cos(\pi/4)} = \sqrt{2} + 1$$

and

$$\dfrac{1 + \sin(\pi/4)\cos(\pi/4)}{\cos(\pi/4)} = \dfrac{1 + (\sqrt{2}/2)(\sqrt{2}/2)}{\sqrt{2}/2} =$$

$$\dfrac{1 + (1/2)}{\sqrt{2}/2} = \dfrac{3}{2} \cdot \dfrac{2}{\sqrt{2}} = \dfrac{3}{\sqrt{2}}.$$

7. True, $\dfrac{1 + \sin x}{1 - \sin^2 x} = \dfrac{1 + \sin x}{(1 - \sin x)(1 + \sin x)} = \dfrac{1}{1 - \sin x}$.

8. True, since $\tan x \cdot \cot x = \tan x \cdot \dfrac{1}{\tan x} = 1$.

9. False, if $x = \pi/3$ then $(1 - \cos(\pi/3))^2 = (1 - 1/2)^2 = (1/2)^2 = 1/4$ and $\sin^2(\pi/3) = (\sqrt{3}/2)^2 = 3/4$.

10. False, if $x = \pi/6$ then $(1 - \csc(\pi/6))(1 + \csc(\pi/6)) = (1 - 2)(1 + 2) = -3$ and $\cot^2(\pi/6) = (\sqrt{3})^2 = 3$.

3.2 Exercises

1. complicated

3. numerator, denominator

5. D, since $\cos x \tan x = \cos x \cdot \dfrac{\sin x}{\cos x} = \sin x$.

7. A, since $\csc^2 x - \cot^2 x = 1$.

9. B, for $1 - \sec^2 x = -\tan^2 x$.

11. H, since $\dfrac{\csc x}{\csc x} - \dfrac{\sin x}{\csc x} = 1 - \sin^2 x = \cos^2 x$.

13. G, for $\csc^2 x = 1 + \cot^2 x$.

15. $\sin^2(\alpha) - 1 = -\cos^2 \alpha$

17. $2\cos^2 \beta - \cos \beta - 1$

19. $\csc^2 x + 2\csc x \sin x + \sin^2 x = \csc^2 x + 2 + \sin^2 x$

21. $4\sin^2 \theta - 1$

23. $9\sin^2 \theta + 12\sin \theta + 4$

25. $4\sin^4 y - 4\sin^2 y \csc^2 y + \csc^4 y = 4\sin^4 y - 4 + \csc^4 y$

27. $(2\sin \gamma + 1)(\sin \gamma - 3)$

29. $(\tan \alpha - 4)(\tan \alpha - 2)$

31. $(2\sec \beta + 1)^2$

33. $(\tan \alpha - \sec \beta)(\tan \alpha + \sec \beta)$

35. $\cos \beta (\sin^2 \beta + \sin \beta - 2) = \cos \beta (\sin \beta + 2)(\sin \beta - 1)$

37. $\left(2\sec^2 x - 1\right)^2$

39. $\cos\alpha(\sin\alpha + 1) + (\sin\alpha + 1) =$
$(\cos\alpha + 1)(\sin\alpha + 1)$

41. Rewrite the left side of the equation.

$$\sin(x)\cot(x) \;=$$

$$\sin(x)\cdot\frac{\cos(x)}{\sin(x)} \;=$$

$$\cos x$$

43. Rewrite the left side of the equation.

$$1 - \sec(x)\cos^3(x) \;=$$

$$1 - \frac{1}{\cos(x)}\cdot\cos^3(x) \;=$$

$$1 - \cos^2(x) \;=$$

$$\sin^2(x)$$

45.

$$1 + \sec^2(x)\sin^2(x) \;=$$

$$1 + \frac{1}{\cos^2(x)}\sin^2(x) \;=$$

$$1 + \frac{\sin^2(x)}{\cos^2(x)} \;=$$

$$1 + \tan^2(x) \;=$$

$$\sec^2(x)$$

47.

$$\frac{\sin^3(x) + \sin(x)\cos^2(x)}{\cos(x)} \;=$$

$$\frac{\sin(x)[\sin^2(x) + \cos^2(x)]}{\cos(x)} \;=$$

$$\frac{\sin(x)[1]}{\cos(x)} \;=$$

$$\tan(x)$$

49.

$$\frac{\sin(x)}{\csc(x)} + \frac{\cos(x)}{\sec(x)} \;=$$

$$\frac{\sin(x)}{1/\sin(x)} + \frac{\cos(x)}{1/\cos(x)} \;=$$

$$\sin^2(x) + \cos^2(x) \;=$$

$$1$$

51. Rewrite the left side of the equation.

$$\tan(x)\cos(x) + \csc(x)\sin^2(x) \;=$$

$$\sin x + \sin x \;=$$

$$2\sin x$$

53.

$$(1 + \sin\alpha)^2 + \cos^2\alpha \;=$$

$$1 + 2\sin\alpha + \sin^2\alpha + \cos^2\alpha \;=$$

$$2 + 2\sin\alpha$$

55.

$$2 - \csc(\beta)\sin(\beta) \;=$$

$$2 - 1 \;=$$

$$1 \;=$$

$$\sin^2(\beta) + \cos^2(\beta)$$

57.

$$\tan x + \cot x \;=$$

$$\frac{\sin x}{\cos x} + \frac{\cos x}{\sin x} \;=$$

$$\frac{\sin^2 x + \cos^2 x}{\sin(x)\cos(x)} \;=$$

$$\frac{1}{\sin(x)\cos(x)} \;=$$

$$\sec(x)\csc(x)$$

59.

$$\frac{\sec(x)}{\tan(x)} - \frac{\tan(x)}{\sec(x)} \;=$$

$$\frac{\sec^2(x) - \tan^2(x)}{\tan(x)\sec(x)} \;=$$

$$\frac{1}{\tan(x)\sec(x)} \;=$$

$$\cot(x)\cos(x)$$

61. Rewrite the right side of the equation.

$$\frac{\csc x}{\csc x - \sin x}$$

$$= \frac{\csc x}{\csc x - \sin x} \cdot \frac{\sin x}{\sin x}$$

$$= \frac{1}{1 - \sin^2 x}$$

$$= \frac{1}{\cos^2 x}$$

$$= \sec^2 x$$

63. Rewrite the right side of the equation.

$$\frac{\cos(-x) - \csc(-x)}{\cos x}$$

$$= \frac{\cos x + \csc x}{\cos x}$$

$$= \frac{\cos x}{\cos x} + \frac{\csc x}{\cos x}$$

$$= 1 + \csc x \sec x$$

65. Rewrite the left side of the equation.

$$\frac{1}{\csc \theta - \cot \theta} =$$

$$\frac{1}{\csc \theta - \cot \theta} \cdot \frac{\sin \theta}{\sin \theta} =$$

$$\frac{\sin \theta}{1 - \cos \theta} =$$

$$\frac{\sin \theta}{1 - \cos \theta} \cdot \frac{1 + \cos \theta}{1 + \cos \theta} =$$

$$\frac{\sin \theta (1 + \cos \theta)}{1 - \cos^2 \theta} =$$

$$\frac{\sin \theta (1 + \cos \theta)}{\sin^2 \theta} =$$

$$\frac{1 + \cos \theta}{\sin \theta}$$

67. Rewrite the right side of the equation.

$$= \frac{1 + \sin(y)}{1 - \sin(y)}$$

$$= \frac{1 + \sin(y)}{1 - \sin(y)} \cdot \frac{\csc(y)}{\csc(y)}$$

$$= \frac{\csc(y) + 1}{\csc(y) - 1}$$

69. Rewrite the left side of the equation.

$$\frac{\cot x + \tan x}{\csc x} =$$

$$\frac{\dfrac{\cos x}{\sin x} + \dfrac{\sin x}{\cos x}}{\dfrac{1}{\sin x}} =$$

$$\frac{\dfrac{\cos^2 x + \sin^2 x}{\sin x \cos x}}{\dfrac{1}{\sin x}} =$$

$$\frac{\dfrac{1}{\sin x \cos x}}{\dfrac{1}{\sin x}} =$$

$$\frac{1}{\cos x}$$

71. Rewrite the left side of the equation.

$$\frac{1 - \sin(-x))^2}{1 - \sin(-x)} =$$

$$\frac{1 - (-\sin x)^2}{1 + \sin x} =$$

$$\frac{1 - \sin^2 x}{1 + \sin x} =$$

$$\frac{(1 - \sin x)(1 + \sin x)}{1 + \sin x} =$$

$$1 - \sin(x)$$

73. Rewrite the left side of the equation.

$$\frac{1 - \cot^2 w + \cos^2 w \cot^2 w}{\csc^2 w} =$$

$$\frac{1 - \cot^2 w (1 - \cos^2 w)}{\csc^2 w} =$$

$$\frac{1 - \cot^2 w \sin^2 w}{\csc^2 w} =$$

$$\frac{1 - \dfrac{\cos^2 w}{\sin^2 w} \sin^2 w}{\csc^2 w} =$$

$$\frac{1 - \cos^2 w}{\csc^2 w} =$$

$$\frac{\sin^2 w}{\csc^2 w} =$$

$$\sin^4 w$$

75.

$$\ln(\sec\theta) =$$
$$\ln((\cos\theta)^{-1}) =$$
$$-\ln(\cos\theta)$$

77. Rewrite the left side of the equation.

$$\ln|\sec\alpha + \tan\alpha| =$$
$$\ln\left|(\sec\alpha + \tan\alpha)\cdot\frac{\sec\alpha - \tan\alpha}{\sec\alpha - \tan\alpha}\right| =$$
$$\ln\left|\frac{\sec^2\alpha - \tan^2\alpha}{\sec\alpha - \tan\alpha}\right| =$$
$$\ln\left|\frac{1}{\sec\alpha - \tan\alpha}\right| =$$
$$\ln\left|(\sec\alpha - \tan\alpha)^{-1}\right| =$$
$$-\ln|\sec\alpha - \tan\alpha|$$

79. It is an identity since

$$\frac{\sin\theta}{\sin\theta} + \frac{\cos\theta}{\sin\theta} =$$
$$1 + \cot\theta.$$

The graphs of $y = \dfrac{\sin\theta + \cos\theta}{\sin\theta}$ and

$y = 1 + \cot\theta$ are shown to be identical.

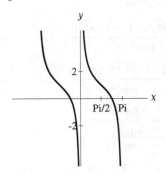

81. It is not an identity since the graphs of

$y = (\sin x + \csc x)^2$ and $y = \sin^2 x + \csc^2 x$

do not coincide as shown.

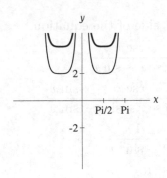

83. It is an identity. Re-arranging the numerator in the right-hand side, we find

$$\frac{1 + \cos x - \cos^2 x}{\sin x}$$
$$= \frac{1 - \cos^2 x + \cos x}{\sin x}$$
$$= \frac{\sin^2 x + \cos x}{\sin x}$$
$$= \frac{\sin^2 x}{\sin x} + \frac{\cos x}{\sin x}$$
$$= \sin x + \cot x.$$

The graphs of $y = \cot x + \sin x$ and

$y = \dfrac{1 + \cos x - \cos^2 x}{\sin x}$ are shown to

be identical.

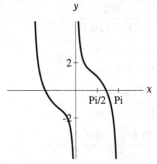

85. It is not an identity since the graphs of

$y = \dfrac{\sin x}{\cos x} - \dfrac{\cos x}{\sin x}$ and $y = \dfrac{2\cos^2 x - 1}{\sin x \cos x}$

are not the same as shown.

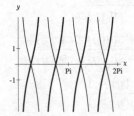

87. It is an identity.

$$\frac{\cos(-x)}{1-\sin(x)} =$$

$$\frac{\cos(x)}{1-\sin(x)} =$$

$$\frac{\cos x}{1-\sin(x)} \cdot \frac{1+\sin x}{1+\sin x} =$$

$$\frac{\cos x(1+\sin x)}{1-\sin^2 x} =$$

$$\frac{\cos x(1+\sin x)}{\cos^2 x} =$$

$$\frac{1+\sin x}{\cos x} =$$

$$\frac{1-\sin(-x)}{\cos x}$$

The graphs of $y = \dfrac{\cos(-x)}{1-\sin x}$ and

$y = \dfrac{1-\sin(-x)}{\cos x}$ are shown to be identical.

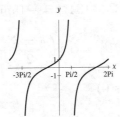

89. For instance, let $f_1(x) = \sin x$ and $f_2(x) = 0$

93. Solve for v_0:

$$\frac{v_0^2 \sin 66^\circ}{32} = 200$$

$$v_0 = \sqrt{\frac{200(32)}{\sin 66^\circ}} \text{ ft/sec}$$

$$v_0 = \sqrt{\frac{200(32)(3600)}{5280 \sin 66^\circ}} \text{ mph}$$

$$v_0 \approx 57.1 \text{ mph}$$

95. The amplitude is $|A| = 4$. Since

$$y = -4\sin\left(\frac{2\pi}{3}\left(x - \frac{1}{2}\right)\right)$$

the period is $\dfrac{2\pi}{B} = \dfrac{2\pi}{2\pi/3} = 3$, and the

phase shift is $C = \dfrac{1}{2}$.

97. $\sin^2 x + \cos^2 x = 1$, $1 + \cot^2 x = \csc^2 x$, and $\tan^2 x + 1 = \sec^2 x$

99. Using a common denominator, we find

$$\frac{19}{40} = \frac{95}{200}, \quad \frac{12}{25} = \frac{96}{200}.$$

All together, there are 201 workers. Then Pat is a male, and Chris is a female.

For Thought

1. False, since $\dfrac{\pi}{4} + \dfrac{\pi}{2} = \dfrac{3\pi}{4}$.

2. False, since $\dfrac{\pi}{4} - \dfrac{\pi}{3} = -\dfrac{\pi}{12}$.

3. False, the right-hand side should be $\cos(5^\circ)$.

4. True, by the sum identity for cosine.

5. True, since $\cos\left(\dfrac{\pi}{2} - \alpha\right) = \sin\alpha$.

6. True, since $\cos(-x) = \cos(x)$.

7. False, rather $\sin(\pi/7 - 3) = -\sin(3 - \pi/7)$.

8. False, since $\sin(x - \pi/2) \neq \cos(x)$ if $x = 0$.

9. True, for $\sec(\pi/3) = \sec(\pi/2 - \pi/6) = \csc(\pi/6)$.

10. True, since the cofunction identity for tangent is applied to $90^\circ - 68^\circ 29' 55'' = 21^\circ 30' 5''$.

3.3 Exercises

1. cosine

3. cotangent

5. $\dfrac{7\pi}{12}$

7. $\dfrac{13\pi}{12}$

9. $\dfrac{3\pi}{12} - \dfrac{4\pi}{12} = -\dfrac{\pi}{12}$

11. $\dfrac{3\pi}{12} - \dfrac{2\pi}{12} = \dfrac{\pi}{12}$

13. $30° + 45°$

15. $120° + 45°$

17. $\dfrac{\pi}{3} - \dfrac{\pi}{4}$

19. $\dfrac{\pi}{3} + \dfrac{\pi}{4}$

21. $\cos(60° - 45°) =$
$\cos(60°)\cos(45°) + \sin(60°)\sin(45°) =$
$\dfrac{1}{2} \cdot \dfrac{\sqrt{2}}{2} + \dfrac{\sqrt{3}}{2} \cdot \dfrac{\sqrt{2}}{2} = \dfrac{\sqrt{2}+\sqrt{6}}{4}$

23. $\cos(60° + 45°) =$
$\cos(60°)\cos(45°) - \sin(60°)\sin(45°) =$
$\dfrac{1}{2} \cdot \dfrac{\sqrt{2}}{2} - \dfrac{\sqrt{3}}{2} \cdot \dfrac{\sqrt{2}}{2} = \dfrac{\sqrt{2}-\sqrt{6}}{4}$

25. $\cos\left(\dfrac{5\pi}{12}\right) = \cos\left(\dfrac{\pi}{6} + \dfrac{\pi}{4}\right) =$
$\cos\left(\dfrac{\pi}{6}\right)\cos\left(\dfrac{\pi}{4}\right) - \sin\left(\dfrac{\pi}{6}\right)\sin\left(\dfrac{\pi}{4}\right) =$
$\dfrac{\sqrt{3}}{2} \cdot \dfrac{\sqrt{2}}{2} - \dfrac{1}{2} \cdot \dfrac{\sqrt{2}}{2} = \dfrac{\sqrt{6}-\sqrt{2}}{4}$

27. $\cos\left(\dfrac{13\pi}{12}\right) = \cos\left(\dfrac{3\pi}{4} + \dfrac{\pi}{3}\right) =$
$\cos\left(\dfrac{3\pi}{4}\right)\cos\left(\dfrac{\pi}{3}\right) - \sin\left(\dfrac{3\pi}{4}\right)\sin\left(\dfrac{\pi}{3}\right) =$
$-\dfrac{\sqrt{2}}{2} \cdot \dfrac{1}{2} - \dfrac{\sqrt{2}}{2} \cdot \dfrac{\sqrt{3}}{2} = -\dfrac{\sqrt{2}+\sqrt{6}}{4}$

29. $\cos\left(-\dfrac{\pi}{12}\right) = \cos\left(\dfrac{\pi}{4} - \dfrac{\pi}{3}\right) =$
$\cos\left(\dfrac{\pi}{4}\right)\cos\left(\dfrac{\pi}{3}\right) + \sin\left(\dfrac{\pi}{4}\right)\sin\left(\dfrac{\pi}{3}\right) =$
$\dfrac{\sqrt{2}}{2} \cdot \dfrac{1}{2} + \dfrac{\sqrt{2}}{2} \cdot \dfrac{\sqrt{3}}{2} = \dfrac{\sqrt{2}+\sqrt{6}}{4}$

31. $\cos\left(-\dfrac{13\pi}{12}\right) = \cos\left(\dfrac{13\pi}{12}\right) =$
$\cos\left(\dfrac{3\pi}{4} + \dfrac{\pi}{3}\right) =$
$\cos\left(\dfrac{3\pi}{4}\right)\cos\left(\dfrac{\pi}{3}\right) - \sin\left(\dfrac{3\pi}{4}\right)\sin\left(\dfrac{\pi}{3}\right) =$
$-\dfrac{\sqrt{2}}{2} \cdot \dfrac{1}{2} - \dfrac{\sqrt{2}}{2} \cdot \dfrac{\sqrt{3}}{2} = -\dfrac{\sqrt{2}+\sqrt{6}}{4}$

33. $\cos(23° - 67°) = \cos(-44°) = \cos(44°)$

35. $\cos(5 + 6) = \cos(11)$

37. $\cos(2k - k) = \cos(k)$

39. $\cos\left(\dfrac{\pi}{2} - \dfrac{\pi}{5}\right) = \cos\left(\dfrac{3\pi}{10}\right)$

41. The problem can be expressed as
$\cos\left(-\dfrac{\pi}{5}\right)\cos\left(-\dfrac{\pi}{3}\right) + \sin\left(-\dfrac{\pi}{5}\right)\sin\left(-\dfrac{\pi}{3}\right) =$
$\cos\left(-\dfrac{\pi}{5} - \left(-\dfrac{\pi}{3}\right)\right) = \cos\left(\dfrac{2\pi}{15}\right)$

43. e, $\sin(20°) = \cos(90° - 20°) = \cos(70°)$

45. c, $\cos(90°) = 0 = \sin(0°)$

47. a, $\sec(\pi/6) = \csc(\pi/2 - \pi/6) = \csc(\pi/3)$

49. g, $\sin(5\pi/12) = \cos(\pi/2 - 5\pi/12) = \cos(\pi/12)$

51. g, $\cos(44°) = \sin(90° - 44°) = \sin(46°)$

53. c, $\cot(134°) = \tan(90° - 134°) = -\tan(44°)$

55. f, $\sec(1) = \csc\left(\dfrac{\pi}{2} - 1\right) = \csc\left(\dfrac{\pi - 2}{2}\right)$

57. a, $\csc(\pi/2) = 1 = \cos(0)$

59. Note, $\cos(61°) = \sin(29°)$. Rewriting we get
$\cos(14°)\cos(29°) + \sin(14°)\sin(29°) =$
$\cos(14° - 29°) = \cos(-15°) = \cos(15°)$.

61. Note, $\cos(-10°) = \cos(10°)$ and
$\cos(70°) = \sin(20°)$. Rewriting we obtain
$\cos(-10°)\cos(20°) + \sin(-10°)\sin(20°) =$
$\cos(-10° - 20°) = \cos(-30°) =$
$\cos(30°) = \dfrac{\sqrt{3}}{2}$.

63. Rewriting we get
$\cos(\pi/2 - \alpha)\cos(-\alpha) - \sin(\alpha - \pi/2)\sin(-\alpha) =$
$\cos(\pi/2 - \alpha)\cos(-\alpha) + \sin(\pi/2 - \alpha)\sin(-\alpha) =$
$\cos((\pi/2 - \alpha) - (-\alpha)) = \cos(\pi/2) = 0$.

65. Rewriting we find
$\cos(-3k)\cos(-k) - \cos(\pi/2 - 3k)\sin(-k) =$
$\cos(3k)\cos(-k) - \sin(3k)\sin(-k) =$
$\cos(3k + (-k)) = \cos(2k)$.

67. Since α is in quadrant II and β is in quadrant I, we find

$$\cos\alpha = -\sqrt{1-\left(\frac{3}{5}\right)^2} = -\sqrt{1-\frac{9}{25}} =$$

$$-\sqrt{\frac{16}{25}} = -\frac{4}{5} \text{ and } \cos\beta = \sqrt{1-\left(\frac{5}{13}\right)^2} =$$

$$\sqrt{1-\frac{25}{169}} = \sqrt{\frac{144}{169}} = \frac{12}{13}. \text{ Then}$$

$$\cos(\alpha+\beta) = \cos\alpha\cos\beta - \sin\alpha\sin\beta =$$

$$-\frac{4}{5}\cdot\frac{12}{13} - \frac{3}{5}\cdot\frac{5}{13} = -\frac{63}{65}.$$

69. Since α is in quadrant IV and β is in quadrant II, we obtain

$$\cos\alpha = \sqrt{1-\left(\frac{-7}{25}\right)^2} = \frac{24}{25}$$

and

$$\cos\beta = -\sqrt{1-\left(\frac{8}{17}\right)^2} = -\frac{15}{17}.$$

Then

$$\cos(\alpha+\beta) = \cos\alpha\cos\beta - \sin\alpha\sin\beta =$$

$$\frac{24}{25}\cdot\left(\frac{-15}{17}\right) - \left(\frac{-7}{25}\right)\cdot\frac{8}{17} = -\frac{304}{425}.$$

71. Since α is in quadrant I and β is in quadrant II, we find

$$\cos\alpha = \sqrt{1-\left(\frac{\sqrt{3}}{2}\right)^2} = \frac{1}{2}$$

and

$$\sin\beta = \sqrt{1-\left(\frac{-\sqrt{2}}{2}\right)^2} = \frac{\sqrt{2}}{2}.$$

Then

$$\cos(\alpha-\beta) = \cos\alpha\cos\beta + \sin\alpha\sin\beta =$$

$$\frac{1}{2}\cdot\left(\frac{-\sqrt{2}}{2}\right) + \frac{\sqrt{3}}{2}\cdot\frac{\sqrt{2}}{2} = \frac{\sqrt{6}-\sqrt{2}}{4}.$$

73. Since α is in quadrant I and β is in quadrant III, we obtain

$$\cos\alpha = \sqrt{1-\left(\frac{2}{3}\right)^2} = \sqrt{1-\frac{4}{9}} =$$

$$\sqrt{\frac{5}{9}} = \frac{\sqrt{5}}{3} \text{ and } \cos\beta = -\sqrt{1-\left(\frac{-1}{2}\right)^2} =$$

$$-\sqrt{1-\frac{1}{4}} = -\sqrt{\frac{3}{4}} = -\frac{\sqrt{3}}{2}. \text{ Thus,}$$

$$\cos(\alpha+\beta) = \cos\alpha\cos\beta - \sin\alpha\sin\beta =$$

$$\frac{\sqrt{5}}{3}\cdot\frac{-\sqrt{3}}{2} - \frac{2}{3}\cdot\frac{-1}{2} = \frac{2-\sqrt{15}}{6}.$$

75. $\cos(\pi/2-(-\alpha)) = \sin(-\alpha) = -\sin\alpha$

77. $\cos 180°\cos\alpha + \sin 180°\sin\alpha =$
$(-1)\cdot\cos\alpha + 0\cdot\sin\alpha = -\cos\alpha$

79. $\cos(3\pi/2)\cos\alpha + \sin(3\pi/2)\sin\alpha =$
$0\cdot\cos\alpha + (-1)\cdot\sin\alpha = -\sin\alpha$

81. $\cos(90°-(-\alpha)) = \sin(-\alpha) = -\sin(\alpha)$

83. Rewrite left side.

$$
\begin{aligned}
\cos(x-\pi/2) &= \\
\cos x\cos(\pi/2) + \sin x\sin(\pi/2) &= \\
\cos x\cdot 0 + \sin x\cdot 1 &= \\
\sin x &= \\
\cos x\tan x
\end{aligned}
$$

85. Substitute the sum and difference cosine identities into the left-hand side to get a difference of two squares.

$$
\begin{aligned}
\cos(\alpha+\beta)\cos(\alpha-\beta) &= \\
(\cos\alpha\cos\beta)^2 - (\sin\alpha\sin\beta)^2 &= \\
(1-\sin^2\alpha)\cos^2\beta - \sin^2\alpha(1-\cos^2\beta) &= \\
\cos^2\beta - \sin^2\alpha\cos^2\beta - \sin^2\alpha + \sin^2\alpha\cos^2\beta &= \\
\cos^2\beta - \sin^2\alpha
\end{aligned}
$$

87.

$$
\begin{aligned}
\cos(x-y) + \cos(y-x) &= \\
\cos(x-y) + \cos(x-y) &= \\
2\cos(x-y) &= \\
2(\cos x\cos y + \sin x\sin y) &= \\
2\cos x\cos y - 2\sin x\sin y
\end{aligned}
$$

89. In the proof, multiply each te rm by $\cos(\alpha - \beta)$. Also, the sum and difference identities for cosine expresses $\cos(\alpha + \beta)\cos(\alpha - \beta)$ as a difference of two squares.

$$\frac{\cos(\alpha + \beta)}{\cos\alpha + \sin\beta} =$$

$$\frac{\cos(\alpha + \beta)}{\cos\alpha + \sin\beta} \cdot \frac{\cos(\alpha - \beta)}{\cos(\alpha - \beta)} =$$

$$\frac{\cos^2\alpha\cos^2\beta - \sin^2\alpha\sin^2\beta}{(\cos\alpha + \sin\beta)\cos(\alpha - \beta)} =$$

$$\frac{\cos^2\alpha(1 - \sin^2\beta) - (1 - \cos^2\alpha)\sin^2\beta}{(\cos\alpha + \sin\beta)\cos(\alpha - \beta)} =$$

$$\frac{\cos^2\alpha - \cos^2\alpha\sin^2\beta - \sin^2\beta + \cos^2\alpha\sin^2\beta}{(\cos\alpha + \sin\beta)\cos(\alpha - \beta)} =$$

$$\frac{\cos^2\alpha - \sin^2\beta}{(\cos\alpha + \sin\beta)\cos(\alpha - \beta)} =$$

$$\frac{(\cos\alpha - \sin\beta)(\cos\alpha + \sin\beta)}{(\cos\alpha + \sin\beta)\cos(\alpha - \beta)} =$$

$$\frac{\cos\alpha - \sin\beta}{\cos(\alpha - \beta)} =$$

$$\frac{\cos\alpha - \sin\beta}{\cos(\beta - \alpha)}$$

95. $\dfrac{1/\sin x}{1/\cos x} = \dfrac{\cos x}{\sin x} = \cot x$

97. $1 - \sin^2\alpha = \cos^2\alpha$

99. The period is $\dfrac{\pi}{B} = \dfrac{\pi}{2}$.

Solve $2x = k\pi$ where k is any integer.

Then the asymptotes are the vertical lines

$$x = \frac{k\pi}{2}.$$

101. Given below are two circles with radii a. Consider the triangle with a vertex at the top point of intersection and with the centers of the circles as the other two vertices. This triangle is an equilateral triangle.

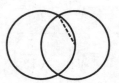

By subtracting the area of a triangle from the area of a sector, we find that the area of the shaded region above is

$$\frac{1}{2}a^2\frac{\pi}{3} - \frac{1}{2}a^2\frac{\sqrt{3}}{2}$$

or

$$\frac{a^2(2\pi - 3\sqrt{3})}{12}.$$

Then the area of the region inside the circle on the right that is outside the circle on the left is

$$2\left[\frac{1}{2}a^2\frac{2\pi}{3}\right] - 2\left[\frac{a^2(2\pi - 3\sqrt{3})}{12}\right]$$

or equivalently

$$\frac{a^2(2\pi + 3\sqrt{3})}{6}.$$

If we add the area of the circle on the left to the above expression we obtain the total area sprinkled, that is,

$$\pi a^2 + \frac{a^2(2\pi + 3\sqrt{3})}{6} = \frac{a^2(8\pi + 3\sqrt{3})}{6}.$$

For Thought

1. True, by the difference identity for sine.

2. False, since the left side of the equation is $\sin 9$.

3. True, by the sum identity for sine.

4. False, since the equation is false if $\alpha = \beta = \dfrac{\pi}{4}$.

5. True, since $\sin 2\pi = 0 = \sin\pi$.

6. False, since $\sin(\pi/2) = 1$ and $\sin(\pi/4) + \sin(\pi/4) = \sqrt{2}$.

7. False, since $\tan 10° \approx 0.176$ and $\tan(2°) + \tan(8°) \approx 0.175$.

8. True, by the sum identity for tangent.

9. True, by the difference identity for tangent.

10. True, since both sides of the equation (by the sum identity for tangent) are equal to $\tan(-7°)$.

3.4 Exercises

1. $\dfrac{3\pi}{12} + \dfrac{\pi}{12} = \dfrac{4\pi}{12} = \dfrac{\pi}{3}$

3. $\dfrac{9\pi}{12} - \dfrac{4\pi}{12} = \dfrac{5\pi}{12}$

5. $60° + 45°$

7. $30° - 45°$

9. $\sin(\pi/3 + \pi/4) =$
$\sin(\pi/3)\cos(\pi/4) + \cos(\pi/3)\sin(\pi/4) =$
$\dfrac{\sqrt{3}}{2} \cdot \dfrac{\sqrt{2}}{2} + \dfrac{1}{2} \cdot \dfrac{\sqrt{2}}{2} = \dfrac{\sqrt{6} + \sqrt{2}}{4}$

11. $\tan(45° + 30°) = \dfrac{\tan(45°) + \tan(30°)}{1 - \tan(45°)\tan(30°)} =$
$\dfrac{1 + \sqrt{3}/3}{1 - 1 \cdot \sqrt{3}/3} \cdot \dfrac{3}{3} = \dfrac{3 + \sqrt{3}}{3 - \sqrt{3}} \cdot \dfrac{3 + \sqrt{3}}{3 + \sqrt{3}} =$
$\dfrac{12 + 6\sqrt{3}}{9 - 3} = 2 + \sqrt{3}$

13. $\sin(30° - 45°) =$
$\sin(30°)\cos(45°) - \cos(30°)\sin(45°) =$
$\dfrac{1}{2} \cdot \dfrac{\sqrt{2}}{2} - \dfrac{\sqrt{3}}{2} \cdot \dfrac{\sqrt{2}}{2} = \dfrac{\sqrt{2} - \sqrt{6}}{4}$

15. $\tan(-13\pi/12) = -\tan(13\pi/12) =$
$-\tan\left(\dfrac{3\pi}{4} + \dfrac{\pi}{3}\right) =$
$-\dfrac{\tan(3\pi/4) + \tan(\pi/3)}{1 - \tan(3\pi/4)\tan(\pi/3)} =$
$-\dfrac{-1 + \sqrt{3}}{1 - (-1)\sqrt{3}} = \dfrac{1 - \sqrt{3}}{1 + \sqrt{3}} \cdot \dfrac{1 - \sqrt{3}}{1 - \sqrt{3}} =$
$\dfrac{4 - 2\sqrt{3}}{-2} = -2 + \sqrt{3}$

17. $\sin(23° + 67°) = \sin(90°) = 1$

19. $\sin(34°)\cos(13°) - \cos(34°)\sin(13°) =$
$\sin(34° - 13°) = \sin(21°)$

21. $\sin\left(-\dfrac{\pi}{2}\right)\cos\left(-\dfrac{\pi}{5}\right) + \cos\left(-\dfrac{\pi}{2}\right)\sin\left(-\dfrac{\pi}{5}\right) =$
$\sin\left(-\dfrac{\pi}{2} + \left(-\dfrac{\pi}{5}\right)\right) = \sin\left(-\dfrac{7\pi}{10}\right) =$
$-\sin\left(\dfrac{7\pi}{10}\right)$

23. $\sin(14°)\cos(35°) + \cos(14°)\sin(35°) =$
$\sin(14° + 35°) = \sin(49°)$

25. $\tan\left(\dfrac{\pi}{9} + \dfrac{\pi}{6}\right) = \tan\left(\dfrac{5\pi}{18}\right)$

27. $\dfrac{\tan(\pi/7) + \tan(\pi/6)}{1 - \tan(\pi/7)\tan(\pi/6)} = \tan\left(\dfrac{\pi}{7} + \dfrac{\pi}{6}\right) =$
$\tan(13\pi/42)$

29. Since α is in quadrant II and β is in quadrant I, we obtain
$$\cos\alpha = -\sqrt{1 - \left(\dfrac{3}{5}\right)^2} = -\sqrt{1 - \dfrac{9}{25}} =$$
$$-\sqrt{\dfrac{16}{25}} = -\dfrac{4}{5} \text{ and } \cos\beta = \sqrt{1 - \left(\dfrac{5}{13}\right)^2} =$$
$$\sqrt{1 - \dfrac{25}{169}} = \sqrt{\dfrac{144}{169}} = \dfrac{12}{13}. \text{ Then}$$
$\sin(\alpha + \beta) = \sin\alpha\cos\beta + \cos\alpha\sin\beta =$
$\dfrac{3}{5} \cdot \dfrac{12}{13} + \dfrac{-4}{5} \cdot \dfrac{5}{13} = \dfrac{16}{65}.$

31. Since α is in quadrant II and β is in quadrant III, we find
$$\cos\alpha = -\sqrt{1 - \left(\dfrac{7}{25}\right)^2} = -\dfrac{24}{25}$$
and
$$\cos\beta = -\sqrt{1 - \left(\dfrac{-8}{17}\right)^2} = -\dfrac{15}{17}.$$
Then
$\sin(\alpha + \beta) = \sin\alpha\cos\beta + \cos\alpha\sin\beta =$
$\dfrac{7}{25} \cdot \left(\dfrac{-15}{17}\right) + \dfrac{-24}{25} \cdot \left(\dfrac{-8}{17}\right) = \dfrac{87}{425}.$

33. Since α is in quadrant II and β is in quadrant I, we find

$$\cos\alpha = -\sqrt{1 - \left(\frac{\sqrt{3}}{2}\right)^2} = -\frac{1}{2}$$

and

$$\sin\beta = \sqrt{1 - \left(\frac{\sqrt{2}}{2}\right)^2} = \frac{\sqrt{2}}{2}.$$

Then

$$\sin(\alpha - \beta) = \sin\alpha\cos\beta - \cos\alpha\sin\beta =$$
$$\frac{\sqrt{3}}{2}\cdot\frac{\sqrt{2}}{2} - \left(\frac{-1}{2}\right)\cdot\frac{\sqrt{2}}{2} = \frac{\sqrt{6}+\sqrt{2}}{4}.$$

35. Since α is in quadrant I and β is in quadrant III, we find

$$\cos\alpha = \sqrt{1 - \left(\frac{2}{3}\right)^2} = \sqrt{1 - \frac{4}{9}} =$$

$$\sqrt{\frac{5}{9}} = \frac{\sqrt{5}}{3} \text{ and } \cos\beta = -\sqrt{1 - \left(\frac{-1}{2}\right)^2} =$$

$$-\sqrt{1 - \frac{1}{4}} = -\sqrt{\frac{3}{4}} = -\frac{\sqrt{3}}{2}. \text{ Thus,}$$

$$\sin(\alpha + \beta) = \sin\alpha\cos\beta + \cos\alpha\sin\beta =$$
$$\frac{2}{3}\cdot\frac{-\sqrt{3}}{2} + \frac{\sqrt{5}}{3}\cdot\frac{-1}{2} = \frac{-2\sqrt{3} - \sqrt{5}}{6}.$$

37. $\sin\alpha\cos\pi - \cos\alpha\sin\pi =$
$\sin\alpha\cdot(-1) - \cos\alpha\cdot 0 = -\sin\alpha$

39. Since the period of $\sin x$ is $360°$, we get $\sin(360° - \alpha) = \sin(-\alpha) = -\sin\alpha.$

41. $\dfrac{\tan(\pi/4) + \tan\alpha}{1 - \tan(\pi/4)\tan\alpha} = \dfrac{1 + \tan\alpha}{1 - 1\cdot\tan\alpha} =$

$$\frac{1 + \tan\alpha}{1 - \tan\alpha}$$

43. Since the period of $y = \tan x$ is π or $180°$, we find $\tan(180° + \alpha) = \tan\alpha.$

45. We rewrite the left side.

$$\begin{aligned}\sin(180° - \alpha) &= \\ \sin(180°)\cos\alpha - \cos(180°)\sin\alpha &= \\ 0\cdot\cos\alpha - (-1)\sin\alpha &= \\ \sin\alpha &= \end{aligned}$$

$$\frac{\sin^2\alpha}{\sin\alpha} =$$

$$\frac{1 - \cos^2\alpha}{\sin\alpha}$$

47. Substitute the sum and difference identities for sine. Then you obtain a difference of two squares as shown:

$$\begin{aligned}\sin(\alpha + \beta)\sin(\alpha - \beta) &= \\ (\sin\alpha\cos\beta)^2 - (\cos\alpha\sin\beta)^2 &= \\ \sin^2\alpha(1 - \sin^2\beta) - (1 - \sin^2\alpha)\sin^2\beta &= \\ \sin^2\alpha - \sin^2\alpha\sin^2\beta - \sin^2\beta + \sin^2\alpha\sin^2\beta &= \\ \sin^2\alpha - \sin^2\beta \end{aligned}$$

49. Rewrite the left side of the identity.

$$\begin{aligned}\sin(x - y) - \sin(y - x) &= \\ \sin(x - y) + \sin(x - y) &= \\ 2\sin(x - y) &= \\ 2(\sin x\cos y - \cos x\sin y) &= \\ 2\sin x\cos y - 2\cos x\sin y \end{aligned}$$

51. Apply the cofunction identity for cotangent.

$$\begin{aligned}\tan(\pi/4 + x) &= \\ \cot(\pi/2 - (\pi/4 + x)) &= \\ \cot(\pi/2 - \pi/4 - x) &= \\ \cot(\pi/4 - x) \end{aligned}$$

53. Use sum and difference identities, then divide each term by $\cos\alpha\cos\beta$.

$$\frac{\cos(\alpha - \beta)}{\sin(\alpha + \beta)} =$$

$$\frac{\dfrac{\cos\alpha\cos\beta}{\cos\alpha\cos\beta} + \dfrac{\sin\alpha\sin\beta}{\cos\alpha\cos\beta}}{\dfrac{\sin\alpha\cos\beta}{\cos\alpha\cos\beta} + \dfrac{\cos\alpha\sin\beta}{\cos\alpha\cos\beta}} =$$

$$\frac{1 + \tan(\alpha)\tan(\beta)}{\tan(\alpha) + \tan(\beta)}$$

55. In the proof, divide the numerator and denominator by $\sin x \sin y$.

$$\frac{\sin(x+y)}{\sin(x-y)} =$$

$$\frac{\sin(x)\cos(y) + \cos(x)\sin(y)}{\sin(x)\cos(y) - \cos(x)\sin(y)} =$$

$$\frac{\dfrac{\sin(x)\cos(y)}{\sin(x)\sin(y)} + \dfrac{\cos(x)\sin(y)}{\sin(x)\sin(y)}}{\dfrac{\sin(x)\cos(y)}{\sin(x)\sin(y)} - \dfrac{\cos(x)\sin(y)}{\sin(x)\sin(y)}} =$$

$$\frac{\cot(y) + \cot(x)}{\cot(y) - \cot(x)}$$

59. a) $-\sin x$ b) $\cos x$ c) $-\tan x$

d) $-\csc x$ e) $\sec x$ f) $\cot x$

61. $\dfrac{\cos x(\cos^2 x + \sin^2 x)}{\sin x} = \dfrac{\cos x}{\sin x} = \cot x$

63. The legs are $38.6 \sin 26° \approx 16.9$ in. and $38.6 \cos 26° \approx 34.7$ in.

65. The angles in the shaded triangle are 15°, 60°, and 105°. The side opposite the 105°-angle is one unit 1 long. Let x be the side AF. To simplify the calculations, we use the sine law (see Chapter 5), we find

$$x = \frac{\sin 60°}{\sin 105°}.$$

Recall, the area of a triangle is one-half times the product of the length of any two sides and the sine of the included angle of the two sides. Since the included angle between x and 1 is 15°, the area of the triangle is

$$\begin{aligned}
\text{Area} &= \frac{1}{2}\frac{\sin 60°}{\sin 105°}\sin 15° \\
&= \frac{\sqrt{3}}{4}\frac{\sin 15°}{\sin 105°} \\
&= \frac{\sqrt{3}}{4}\frac{\sqrt{(2-\sqrt{3}/2)/2}}{\sqrt{(2+\sqrt{3}/2)/2}} \\
&= \frac{\sqrt{3}}{4}\frac{\sqrt{2-\sqrt{3}}}{\sqrt{2+\sqrt{3}}}
\end{aligned}$$

$$= \frac{\sqrt{3}}{4}(2 - \sqrt{3})$$

$$\text{Area} = \frac{2\sqrt{3} - 3}{4}.$$

For Thought

1. True, $\dfrac{\sin(2 \cdot 21°)}{2} = \dfrac{2\sin(21°)\cos(21°)}{2}$

$= \sin(21°)\cos(21°).$

2. True, by a cosine double angle identity

$$\cos(2\sqrt{2}) = 2\cos^2(\sqrt{2}) - 1.$$

3. False, $\sin\left(\dfrac{300°}{2}\right) = \sqrt{\dfrac{1 - \cos(300°)}{2}}.$

4. True, $\sin\left(\dfrac{400°}{2}\right) = -\sqrt{\dfrac{1 - \cos(400°)}{2}}$

$= -\sqrt{\dfrac{1 - \cos(40°)}{2}}.$

5. False, $\tan\left(\dfrac{7\pi/4}{2}\right) = -\sqrt{\dfrac{1 - \cos(7\pi/4)}{1 + \cos(7\pi/4)}}.$

6. True, $\tan\left(\dfrac{-\pi/4}{2}\right) = \dfrac{1 - \cos(-\pi/4)}{\sin(-\pi/4)} =$

$\dfrac{1 - \cos(\pi/4)}{\sin(-\pi/4)}$

7. False, if $x = \pi/4$ then $\dfrac{\sin(2 \cdot \pi/4)}{2} =$

$\dfrac{\sin(\pi/2)}{2} = \dfrac{1}{2}$ and $\sin(\pi/4) = \sqrt{2}/2.$

8. False, since $\cos(2\pi/3) = -1/2$ while

$\sqrt{\dfrac{1 + \cos(2x)}{2}}$ is a non-negative number.

9. True, since $1 - \cos x \geq 0$ we find

$$\sqrt{(1 - \cos x)^2} = |1 - \cos x| = 1 - \cos x.$$

10. True, α is in quadrant III or IV while, $\alpha/2$ is in quadrant II.

3.5 Exercises

1. $\sin(2 \cdot 45°) = 2\sin(45°)\cos(45°) =$

$2 \cdot \dfrac{\sqrt{2}}{2} \cdot \dfrac{\sqrt{2}}{2} = 2 \cdot \dfrac{2}{4} = 1.$

3. $\tan(2 \cdot 30°) = \dfrac{2\tan(30°)}{1 - \tan^2(30°)} = \dfrac{2(\sqrt{3}/3)}{1 - (\sqrt{3}/3)^2} =$

$\dfrac{2\sqrt{3}/3}{1 - 1/3} = \dfrac{2\sqrt{3}/3}{2/3} = \sqrt{3}$

5. $\sin\left(2 \cdot \dfrac{3\pi}{4}\right) = 2\sin(3\pi/4)\cos(3\pi/4) =$

$2 \cdot \dfrac{\sqrt{2}}{2} \cdot \dfrac{-\sqrt{2}}{2} = 2 \cdot \dfrac{-2}{4} = -1$

7. $\tan\left(2 \cdot \dfrac{2\pi}{3}\right) = \dfrac{2\tan(2\pi/3)}{1 - \tan^2(2\pi/3)} = \dfrac{2(-\sqrt{3})}{1 - (-\sqrt{3})^2}$

$= \dfrac{-2\sqrt{3}}{1 - 3} = \dfrac{-2\sqrt{3}}{-2} = \sqrt{3}$

9. $\sin\left(\dfrac{30°}{2}\right) = \sqrt{\dfrac{1 - \cos(30°)}{2}} =$

$\sqrt{\dfrac{1 - \sqrt{3}/2}{2} \cdot \dfrac{2}{2}} = \sqrt{\dfrac{2 - \sqrt{3}}{4}} = \dfrac{\sqrt{2 - \sqrt{3}}}{2}$

11. $\tan\left(\dfrac{30°}{2}\right) = \dfrac{1 - \cos(30°)}{\sin(30°)} =$

$\dfrac{1 - \sqrt{3}/2}{1/2} \cdot \dfrac{2}{2} = 2 - \sqrt{3}$

13. $\cos\left(\dfrac{\pi/4}{2}\right) = \sqrt{\dfrac{1 + \cos(\pi/4)}{2}} =$

$\sqrt{\dfrac{1 + \sqrt{2}/2}{2} \cdot \dfrac{2}{2}} = \sqrt{\dfrac{2 + \sqrt{2}}{4}} = \dfrac{\sqrt{2 + \sqrt{2}}}{2}$

15. $\sin\left(\dfrac{45°}{2}\right) = \sqrt{\dfrac{1 - \cos(45°)}{2}} =$

$\sqrt{\dfrac{1 - \sqrt{2}/2}{2} \cdot \dfrac{2}{2}} = \sqrt{\dfrac{2 - \sqrt{2}}{4}} = \dfrac{\sqrt{2 - \sqrt{2}}}{2}$

17. $\cos(7\pi/8) = -\sqrt{\dfrac{1 + \cos(7\pi/4)}{2}} =$

$-\sqrt{\dfrac{1 + \dfrac{\sqrt{2}}{2}}{2}} = -\sqrt{\dfrac{2 + \sqrt{2}}{4}} = -\dfrac{\sqrt{2 + \sqrt{2}}}{2}$

19. Positive, $118.5°$ is in quadrant II

21. Negative, $100°$ is in quadrant II

23. Negative, $-5\pi/12$ is in quadrant IV

25. $\sin(2 \cdot 13°) = \sin 26°$

27. $\cos(2 \cdot 22.5°) = \cos 45° = \dfrac{\sqrt{2}}{2}$

29. $\dfrac{1}{2} \cdot \dfrac{2\tan 15°}{1 - \tan^2 15°} = \dfrac{1}{2} \cdot \tan(2 \cdot 15°) =$

$\dfrac{1}{2} \cdot \tan 30° = \dfrac{1}{2} \cdot \dfrac{\sqrt{3}}{3} = \dfrac{\sqrt{3}}{6}$

31. $\dfrac{1}{2} \cdot \dfrac{2\tan 30°}{1 - \tan^2 30°} = \dfrac{1}{2} \cdot \tan(2 \cdot 30°) =$

$\dfrac{1}{2} \cdot \tan 60° = \dfrac{1}{2} \cdot \sqrt{3} = \dfrac{\sqrt{3}}{2}$

33. $2\sin\left(\dfrac{\pi}{9} - \dfrac{\pi}{2}\right)\cos\left(\dfrac{\pi}{9} - \dfrac{\pi}{2}\right) =$

$\sin\left(2 \cdot \left(\dfrac{\pi}{9} - \dfrac{\pi}{2}\right)\right) = \sin\left(\dfrac{2\pi}{9} - \pi\right) =$

$\sin\left(-\dfrac{7\pi}{9}\right) = -\sin\left(\dfrac{7\pi}{9}\right).$

35. $\tan\left(\dfrac{12°}{2}\right) = \tan 6°$

37. Rewrite the left side of the identity.

$$\begin{aligned}
\cos^4 s - \sin^4 s &= \\
(\cos^2 s - \sin^2 s)(\cos^2 s + \sin^2 s) &= \\
\cos(2s) \cdot (1) &= \\
\cos(2s)
\end{aligned}$$

39. Apply double angle identities to the left side of the identity.

$$\begin{aligned}
\dfrac{\sin(4t)}{4} &= \\
\dfrac{2\sin(2t)\cos(2t)}{4} &= \\
\dfrac{2 \cdot 2\sin t \cos t \cdot (\cos^2 t - \sin^2 t)}{4} &= \\
\sin t \cos t(\cos^2 t - \sin^2 t) &= \\
\cos^3 t \sin t - \sin^3 t \cos t
\end{aligned}$$

41.

$$\frac{\cos(2x) + \cos(2y)}{\sin(x) + \cos(y)} =$$

$$\frac{1 - 2\sin^2 x + 2\cos^2 y - 1}{\sin x + \cos y} =$$

$$2\frac{\cos^2 y - \sin^2 x}{\sin x + \cos y} =$$

$$2\frac{(\cos y - \sin x)(\cos y + \sin x)}{\sin x + \cos y} =$$

$$2\cos(y) - 2\sin(x)$$

43.

$$\frac{\cos(2x)}{\sin^2 x} =$$

$$\frac{1 - 2\sin^2 x}{\sin^2 x} =$$

$$\frac{1}{\sin^2 x} - 2 \cdot \frac{\sin^2 x}{\sin^2 x} =$$

$$\csc^2 x - 2$$

45. Rewrite the right side of the identity.

$$\frac{\sin^2 u}{1 + \cos u}$$

$$= \frac{1 - \cos^2 u}{1 + \cos u}$$

$$= \frac{(1 - \cos u)(1 + \cos u)}{1 + \cos u}$$

$$= (1 - \cos u) \cdot \frac{2}{2}$$

$$= 2 \cdot \frac{1 - \cos u}{2}$$

$$= 2\sin^2(u/2)$$

47. Multiply and divide by $\cos x$.

$$\frac{\sec x + \cos x - 2}{\sec x - \cos x}$$

$$= \frac{\sec x + \cos x - 2}{\sec x - \cos x} \cdot \frac{\cos x}{\cos x}$$

$$= \frac{1 + \cos^2 x - 2\cos x}{1 - \cos^2 x}$$

$$= \frac{\cos^2 x - 2\cos x + 1}{1 - \cos^2 x}$$

$$= \frac{(1 - \cos x)^2}{(1 + \cos x)(1 - \cos x)}$$

$$= \frac{1 - \cos x}{1 + \cos x}$$

$$= \tan^2(x/2)$$

49.

$$\frac{1 - \sin^2(x/2)}{1 + \sin^2(x/2)} =$$

$$\frac{1 - \left(\dfrac{1 - \cos x}{2}\right)}{1 + \left(\dfrac{1 - \cos x}{2}\right)} \cdot \frac{2}{2} =$$

$$\frac{2 - (1 - \cos x)}{2 + (1 - \cos x)} =$$

$$\frac{1 + \cos x}{3 - \cos x}$$

51. Since $\cos(2\alpha) = 2\cos^2 \alpha - 1$, we get

$$2\cos^2 \alpha - 1 = \frac{3}{5}$$

$$2\cos^2 \alpha = \frac{8}{5}$$

$$\cos^2 \alpha = \frac{4}{5}$$

$$\cos \alpha = \pm\frac{2}{\sqrt{5}}.$$

But $0° < \alpha < 45°$, so $\cos \alpha = \dfrac{2\sqrt{5}}{5}$ and

$$\sin \alpha = \sqrt{1 - \left(\frac{2}{\sqrt{5}}\right)^2} = \sqrt{1 - \frac{4}{5}} = \sqrt{\frac{1}{5}} =$$

$\dfrac{\sqrt{5}}{5}$. Furthermore, $\sec \alpha = \dfrac{\sqrt{5}}{2}$, $\csc \alpha = \sqrt{5}$, $\tan \alpha = \dfrac{1/\sqrt{5}}{2/\sqrt{5}} = \dfrac{1}{2}$, and $\cot \alpha = 2$.

53. Since $0 < 2\alpha < 90°$, we find

$$\cos(2\alpha) = \sqrt{1 - \left(\frac{5}{13}\right)^2} = \sqrt{\frac{144}{169}} = \frac{12}{13}.$$

Then $\sin(\alpha) = \sqrt{\dfrac{1 - \cos 2\alpha}{2}} = \sqrt{\dfrac{1 - \dfrac{12}{13}}{2}} =$

$\sqrt{\dfrac{1}{26}} = \dfrac{\sqrt{26}}{26},$

$\cos(\alpha) = \sqrt{\dfrac{1 + \cos 2\alpha}{2}} = \sqrt{\dfrac{1 + \dfrac{12}{13}}{2}} =$

$\sqrt{\dfrac{25}{26}} = \dfrac{5\sqrt{26}}{26},$

$\tan \alpha = \dfrac{\sqrt{26}/26}{5\sqrt{26}26} = \dfrac{1}{5},\ \csc \alpha = \sqrt{26},$

$\sec \alpha = \sqrt{\dfrac{26}{25}} = \dfrac{\sqrt{26}}{5},\ \text{and}\ \cot \alpha = 5.$

55. By a half-angle identity, we have

$$\begin{aligned}
\cos(\alpha/2) = -\sqrt{\dfrac{1 + \cos \alpha}{2}} &= -\dfrac{1}{4} \\
\dfrac{1 + \cos \alpha}{2} &= \dfrac{1}{16} \\
1 + \cos \alpha &= \dfrac{1}{8} \\
\cos \alpha &= -\dfrac{7}{8}.
\end{aligned}$$

Since $\pi \le \alpha \le 3\pi/2$, we get

$\sin \alpha = -\sqrt{1 - \left(-\dfrac{7}{8}\right)^2} = -\sqrt{1 - \dfrac{49}{64}} =$

$-\sqrt{\dfrac{15}{64}} = -\dfrac{\sqrt{15}}{8}.$ Furthermore,

$\sec \alpha = -\dfrac{8}{7},\ \csc \alpha = -\dfrac{8}{\sqrt{15}} = -\dfrac{8\sqrt{15}}{15},$

$\tan \alpha = \dfrac{-\sqrt{15}/8}{-7/8} = \dfrac{\sqrt{15}}{7},\ \text{and}$

$\cot \alpha = \dfrac{7}{\sqrt{15}} = \dfrac{7\sqrt{15}}{15}.$

57. By a half-angle identity, we find

$$\begin{aligned}
\sin(\alpha/2) = \sqrt{\dfrac{1 - \cos \alpha}{2}} &= \dfrac{4}{5} \\
\dfrac{1 - \cos \alpha}{2} &= \dfrac{16}{25} \\
1 - \cos \alpha &= \dfrac{32}{25} \\
\cos \alpha &= -\dfrac{7}{25}.
\end{aligned}$$

Since $(\pi/2 + 2k\pi) \le \alpha/2 \le (\pi + 2k\pi)$ for some integer k, we get $(\pi + 4k\pi) \le \alpha \le (2\pi + 4k\pi)$. Then α is in quadrant III because $\cos \alpha < 0$.

Thus, $\sin \alpha = -\sqrt{1 - \left(-\dfrac{7}{25}\right)^2} =$

$-\sqrt{1 - \dfrac{49}{625}} = -\sqrt{\dfrac{576}{625}} = -\dfrac{24}{25}.$

Furthermore, $\sec \alpha = -\dfrac{25}{7},\ \csc \alpha = -\dfrac{25}{24},$

$\tan \alpha = \dfrac{-24/25}{-7/25} = \dfrac{24}{7},\ \text{and}\ \cot \alpha = \dfrac{7}{24}.$

59. Since α lies in quadrant 2, we find

$$\cos \alpha = -\sqrt{1 - \left(\dfrac{3}{5}\right)^2} = -\dfrac{4}{5}.$$

Then

$$\sin(2\alpha) = 2\sin \alpha \cos \alpha = 2\left(\dfrac{3}{5}\right)\left(\dfrac{-4}{5}\right) = -\dfrac{24}{25}.$$

61. Since $\sin \alpha = \dfrac{8}{17}$, we obtain

$$\cos 2\alpha = 1 - 2\sin^2 \alpha = 1 - 2\left(\dfrac{8}{17}\right) = \dfrac{161}{289}.$$

63. Since $\tan \alpha = \dfrac{3}{5}$, we get $\sin \alpha = \dfrac{3}{\sqrt{34}}$ and

$$\cos \alpha = \dfrac{5}{\sqrt{34}}.$$

Applying a half-angle identity, we obtain

$$\begin{aligned}
\tan \dfrac{\alpha}{2} &= \dfrac{\sin \alpha}{1 + \cos \alpha} \\
&= \dfrac{\dfrac{3}{\sqrt{34}}}{1 + \dfrac{5}{\sqrt{34}}} \\
&= \dfrac{3}{5 + \sqrt{34}}
\end{aligned}$$

and since $\tan \dfrac{\alpha}{2} = \dfrac{BD}{5}$ then

$$BD = \dfrac{15}{5 + \sqrt{34}}$$

$$= \frac{15(5 - \sqrt{34})}{25 - 34}$$

$$= \frac{15(\sqrt{34} - 5)}{9}$$

$$BD = \frac{5\sqrt{34} - 25}{3}.$$

65. Since the base of the TV screen is $b = d\cos\alpha$ and its height is $h = d\sin\alpha$, then the area A is given by

$$A = bh$$
$$= (d\cos\alpha)(d\sin\alpha)$$
$$= d^2\cos\alpha\sin\alpha$$
$$A = \frac{d^2}{2}\sin(2\alpha).$$

67. It is not an identity. If $x = \pi/4$, then

$$\sin(2 \cdot \pi/4) = \sin(\pi/2) = 1$$

and

$$2\sin(\pi/4) = 2 \cdot (\sqrt{2}/2) = \sqrt{2}.$$

69. It is not an identity. If $x = 2\pi/3$, then

$$\tan\left(\frac{2\pi/3}{2}\right) = \tan(\pi/3) = \sqrt{3}$$

and

$$\frac{1}{2} \cdot \tan(2\pi/3) = \frac{1}{2} \cdot (-\sqrt{3}).$$

71. It is not an identity. If $x = \pi/2$, then

$$\sin\left(2 \cdot \pi/2\right)\sin\left(\frac{\pi/2}{2}\right) = \sin(\pi)\sin(\pi/4)$$

$$= 0 \cdot \frac{\sqrt{2}}{2} = 0 \text{ and } \sin^2(\pi/2) = 1.$$

73. It is an identity. The proof below uses the double-angle identity for tangent.

$$\cot(x/2) - \tan(x/2) =$$
$$\frac{1}{\tan(x/2)} - \tan(x/2) =$$
$$\frac{1 - \tan^2(x/2)}{\tan(x/2)} =$$
$$2 \cdot \frac{1 - \tan^2(x/2)}{2 \cdot \tan(x/2)} =$$

$$2 \cdot \frac{1}{\tan x} =$$
$$2 \cdot \frac{\cos x}{\sin x} \cdot \frac{\sin x}{\sin x} =$$
$$\frac{2\sin x\cos x}{\sin^2 x} =$$
$$\frac{\sin(2x)}{\sin^2 x}$$

75. Since $\tan\dfrac{x}{2} = \dfrac{1 - \cos x}{\sin x}$ and $1 - \cos x \geq 0$, it follows that $\tan\dfrac{x}{2}$ and $\sin x$ have the same signs provided $\cos x \neq 1$.

79. **a)** $\sin x\cos y + \cos x\sin y$

    b) $\sin x\cos y - \cos x\sin y$

81. $\cos\beta = -\sqrt{1 - (2/3)^2} = -\sqrt{\dfrac{5}{9}} = -\dfrac{\sqrt{5}}{3}$

83. Apply a cofunction identity:

$$\sin(\pi/2 - x) = \cos x = \frac{3}{4}$$

85. An eighth of the region that gets watered by all sprinklers is region R_a below with vertices B, C, and D.

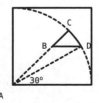

The area of R_a is the area of the sector determined by C, A , and D minus the area of triangle $\triangle ABD$. In the figure above, we have $AB = \sqrt{2}$, $BC = 2 - \sqrt{2}$, $BD = \sqrt{3} - 1$, angle $\langle ABD = 135°$, and $\langle CBD = 45°$.

The area of the sector is

$$A_s = \frac{1}{2}\left(2^2\frac{\pi}{12}\right) = \frac{\pi}{6}$$

and the area of $\triangle ABD$ is

$$A_t = \frac{1}{2}(AB)(BD)\sin 135°$$

$$= \frac{1}{2}\sqrt{2}(\sqrt{3}-1)\frac{\sqrt{2}}{2}$$

$$A_t = \frac{\sqrt{3}-1}{2}.$$

Thus, the area of the region watered by all sprinklers is

$$\text{Area} = 8(A_s - A_t)$$

$$= 8\left(\frac{\pi}{6} - \frac{\sqrt{3}-1}{2}\right)$$

$$\text{Area} = 4\pi/\sqrt{3} + 4 - 4\sqrt{3}.$$

For Thought

1. True, $\sin 45° \cos 15° =$
$(1/2)\left[\sin(45°+15°)+\sin(45°-15°)\right] =$
$0.5\left[\sin 60° + \sin 30°\right].$

2. False, $\cos(\pi/8)\sin(\pi/4) =$
$(1/2)\left[\sin(\pi/8 + \pi/4) - \sin(\pi/8 - \pi/4)\right] =$
$0.5\left[\sin(3\pi/8) - \sin(-\pi/8)\right] =$
$0.5\left[\sin(3\pi/8) + \sin(\pi/8)\right].$

3. True, $2\cos(6°)\cos(8°) =$
$\cos(6° - 8°) + \cos(6° + 8°) =$
$\cos(-2°) + \cos(14°) = \cos(2°) + \cos(14°).$

4. False, $\sin(5°) - \sin(9°) =$
$2\cos\left(\frac{5°+9°}{2}\right)\sin\left(\frac{5°-9°}{2}\right) =$
$2\cos(7°)\sin(-2°) = -2\cos(7°)\sin(2°).$

5. True, $\cos(4) + \cos(12) =$
$2\cos\left(\frac{4+12}{2}\right)\cos\left(\frac{4-12}{2}\right) =$
$2\cos(8)\cos(-4) = 2\cos(8)\cos(4).$

6. False, $\cos(\pi/3) - \cos(\pi/2) =$
$-2\sin\left(\frac{\pi/3 + \pi/2}{2}\right)\sin\left(\frac{\pi/3 - \pi/2}{2}\right) =$
$-2\sin(5\pi/12)\sin(-\pi/12) =$
$2\sin(5\pi/12)\sin(\pi/12).$

7. True, $\sqrt{2}\sin(\pi/6 + \pi/4) =$
$\sqrt{2}\left[\sin(\pi/6)\cos(\pi/4) + \cos(\pi/6)\sin(\pi/4)\right] =$
$\sqrt{2}\left[\sin(\pi/6)\cdot\frac{1}{\sqrt{2}} + \cos(\pi/6)\cdot\frac{1}{\sqrt{2}}\right] =$
$\sin(\pi/6) + \cos(\pi/6).$

8. True, $\frac{1}{2}\sin(\pi/6) + \frac{\sqrt{3}}{2}\cos(\pi/6) =$
$\frac{1}{2}\cdot\frac{1}{2} + \frac{\sqrt{3}}{2}\cdot\frac{\sqrt{3}}{2} = \frac{1}{4} + \frac{3}{4} = 1 = \sin(\pi/2).$

9. True, $y = \cos(\pi/3)\sin x + \sin(\pi/3)\cos x = \sin(x + \pi/3).$

10. True, since $y = \cos(\pi/4)\sin x + \sin(\pi/4)\cos x = \sin(x + \pi/4)$ holds by a sum identity for sine.

3.6 Exercises

1.
$$\frac{1}{2}\left[\cos(13° - 9°) - \cos(13° + 9°)\right] =$$
$$0.5\left[\cos 4° - \cos 22°\right]$$

3.
$$\frac{1}{2}\left[\sin(16° + 20°) + \sin(16° - 20°)\right] =$$
$$0.5\left[\sin 36° + \sin(-4°)\right] = 0.5\left[\sin 36° - \sin 4°\right]$$

5.
$$\frac{1}{2}\left[\cos\left(\frac{\pi}{6} - \frac{\pi}{5}\right) + \cos\left(\frac{\pi}{6} + \frac{\pi}{5}\right)\right] =$$
$$0.5\left[\cos\left(\frac{-\pi}{30}\right) + \cos\left(\frac{11\pi}{30}\right)\right] =$$
$$0.5\left[\cos\left(\frac{\pi}{30}\right) + \cos\left(\frac{11\pi}{30}\right)\right]$$

7.
$$\frac{1}{2}\left[\cos(5y^2 - 7y^2) + \cos(5y^2 + 7y^2)\right] =$$
$$0.5\left[\cos(-2y^2) + \cos(12y^2)\right] =$$
$$0.5\left[\cos(2y^2) + \cos(12y^2)\right]$$

9.
$$\frac{1}{2}\Big[\sin((2s-1) + (s+1)) +$$
$$\sin((2s-1) - (s+1))\Big] =$$
$$0.5\Big[\sin(3s) + \sin(s-2)\Big]$$

11.

$$\frac{1}{2}\left[\cos(52.5° - 7.5°) - \cos(52.5° + 7.5°)\right] =$$

$$\frac{1}{2}\left[\cos 45° - \cos 60°\right] =$$

$$\frac{1}{2}\left[\frac{\sqrt{2}}{2} - \frac{1}{2}\right] = \frac{\sqrt{2} - 1}{4}$$

13.

$$\frac{1}{2}\left[\sin\left(\frac{13\pi}{24} + \frac{5\pi}{24}\right) + \sin\left(\frac{13\pi}{24} - \frac{5\pi}{24}\right)\right] =$$

$$\frac{1}{2}\left[\sin(18\pi/24) + \sin(8\pi/24)\right] =$$

$$\frac{1}{2}\left[\sin(3\pi/4) + \sin(\pi/3)\right] =$$

$$\frac{1}{2}\left[\frac{\sqrt{2}}{2} + \frac{\sqrt{3}}{2}\right] = \frac{\sqrt{2} + \sqrt{3}}{4}$$

15.

$$2\cos\left(\frac{12° + 8°}{2}\right)\sin\left(\frac{12° - 8°}{2}\right) =$$

$$2\cos 10° \sin 2°$$

17.

$$-2\sin\left(\frac{80° + 87°}{2}\right)\sin\left(\frac{80° - 87°}{2}\right) =$$

$$-2\sin 83.5° \sin(-3.5°) = 2\sin 83.5° \sin 3.5°$$

19.

$$2\cos\left(\frac{3.6 + 4.8}{2}\right)\sin\left(\frac{3.6 - 4.8}{2}\right) =$$

$$2\cos(4.2)\sin(-0.6) = -2\cos(4.2)\sin(0.6)$$

21.

$$-2\sin\left(\frac{(5y - 3) + (3y + 9)}{2}\right) \cdot$$

$$\sin\left(\frac{(5y - 3) - (3y + 9)}{2}\right) =$$

$$-2\sin(4y + 3)\sin(y - 6)$$

23.

$$2\cos\left(\frac{5\alpha + 8\alpha}{2}\right)\sin\left(\frac{5\alpha - 8\alpha}{2}\right) =$$

$$2\cos(6.5\alpha)\sin(-1.5\alpha) =$$
$$-2\cos(6.5\alpha)\sin(1.5\alpha)$$

25.

$$-2\sin\left(\frac{\pi/3 + \pi/5}{2}\right)\sin\left(\frac{\pi/3 - \pi/5}{2}\right) =$$

$$-2\sin\left(\frac{4\pi}{15}\right)\sin\left(\frac{\pi}{15}\right)$$

27.

$$2\sin\left(\frac{75° + 15°}{2}\right)\cos\left(\frac{75° - 15°}{2}\right) =$$

$$2\sin 45° \cos(30°) = 2 \cdot \frac{\sqrt{2}}{2}\frac{\sqrt{3}}{2} = \frac{\sqrt{6}}{2}$$

29.

$$-2\sin\left(\frac{\frac{-\pi}{24} + \frac{7\pi}{24}}{2}\right)\sin\left(\frac{\frac{-\pi}{24} - \frac{7\pi}{24}}{2}\right) =$$

$$-2\sin(3\pi/24)\sin(-4\pi/24) =$$
$$-2\sin(\pi/8)\sin(-\pi/6) =$$

$$-2\sin\left(\frac{\pi/4}{2}\right) \cdot \frac{-1}{2} = -2\sqrt{\frac{1 - \cos(\pi/4)}{2}} \cdot \frac{-1}{2}$$

$$= \sqrt{\frac{1 - \sqrt{2}/2}{2} \cdot \frac{2}{2}} = \sqrt{\frac{2 - \sqrt{2}}{4}} = \frac{\sqrt{2 - \sqrt{2}}}{2}$$

31. Since $a = 1$ and $b = -1$, we find

$$r = \sqrt{1^2 + (-1)^2} = \sqrt{2}.$$

If the terminal side of α passes through $(1, -1)$, then $\cos\alpha = a/r = 1/\sqrt{2}$ and $\sin\alpha = b/r = -1/\sqrt{2}$. Let $\alpha = -\pi/4$. Then $\sin x - \cos x = r\sin(x + \alpha) =$

$$\sqrt{2}\sin\left(x - \frac{\pi}{4}\right).$$

33. Since $a = -1/2$ and $b = \sqrt{3}/2$, we obtain

$$r = \sqrt{(-1/2)^2 + (\sqrt{3}/2)^2} = 1.$$

If the terminal side of α passes through $(-1/2, \sqrt{3}/2)$, then $\cos\alpha = a/r = a/1 = a = -1/2$ and $\sin\alpha = b/r = b/1 = b = \sqrt{3}/2$. Let $\alpha = 2\pi/3$. Thus, $-\frac{1}{2}\sin x + \frac{\sqrt{3}}{2}\cos x =$

$$r\sin(x + \alpha) = \sin\left(x + \frac{2\pi}{3}\right).$$

35. Since $a = \sqrt{3}/2$ and $b = -1/2$, we have

$$r = \sqrt{(\sqrt{3}/2)^2 + (-1/2)^2} = 1.$$

If the terminal side of α passes through $(\sqrt{3}/2, -1/2)$, then $\cos\alpha = a/r = a/1 = a = \sqrt{3}/2$ and $\sin\alpha = b/r = b/1 = b = -1/2$. Let $\alpha = -\pi/6$. Thus, $\frac{\sqrt{3}}{2}\sin x - \frac{1}{2}\cos x =$

$$r\sin(x + \alpha) = \sin\left(x - \frac{\pi}{6}\right).$$

37. Since $a = -1$ and $b = 1$, we obtain
$r = \sqrt{(-1)^2 + 1^2} = \sqrt{2}$. If the terminal side
of α passes through $(-1, 1)$, then $\cos \alpha =$
$a/r = -1/\sqrt{2}$ and $\sin \alpha = b/r = 1/\sqrt{2}$.
Choose $\alpha = 3\pi/4$. Then $y = -\sin x + \cos x =$
$r \sin(x + \alpha) = \sqrt{2} \sin(x + 3\pi/4)$. Amplitude
is $\sqrt{2}$, period is 2π, and phase shift is $-3\pi/4$.

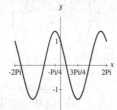

39. Since $a = \sqrt{2}$ and $b = -\sqrt{2}$, we obtain
$r = \sqrt{\sqrt{2}^2 + (-\sqrt{2})^2} = 2$. If the terminal side
of α passes through $(\sqrt{2}, -\sqrt{2})$, then $\cos \alpha =$
$a/r = \sqrt{2}/2$ and $\sin \alpha = b/r = -\sqrt{2}/2$.
Let $\alpha = -\pi/4$.
Thus, $y = \sqrt{2} \sin x - \sqrt{2} \cos x = r \sin(x + \alpha) =$
$2 \sin(x - \pi/4)$. Amplitude is 2, period is 2π,
and phase shift is $\pi/4$.

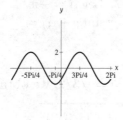

41. Since $a = -\sqrt{3}$ and $b = -1$, we find
$r = \sqrt{(-\sqrt{3})^2 + (-1)^2} = 2$. If the terminal
side of α passes through $(-\sqrt{3}, -1)$,
then $\cos \alpha = a/r = -\sqrt{3}/2$ and
$\sin \alpha = b/r = -1/2$. Choose $\alpha = 7\pi/6$.
Then $y = -\sqrt{3} \sin x - \cos x = r \sin(x + \alpha) =$
$2 \sin(x + 7\pi/6)$. Amplitude is 2, period is 2π,
and phase shift is $-7\pi/6$.

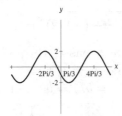

43. Since $a = 3$ and $b = 4$, the amplitude is
$\sqrt{3^2 + 4^2} = \sqrt{25} = 5$. If the terminal side of

α passes through $(3, 4)$, then $\tan \alpha = 3/4$ and
$\alpha = \tan^{-1}(3/4) \approx 0.9$. The phase shift is -0.9.

45. Since $a = -6$ and $b = 1$, the amplitude is
$\sqrt{(-6)^2 + 1^2} = \sqrt{37}$. If the terminal side of α
passes through $(-6, 1)$, then $\tan \alpha = -1/6$.
Using a calculator, one gets $\tan^{-1}(-1/6) \approx$
-0.165 which is an angle in quadrant IV. Since
$(-6, 1)$ is in quadrant II and π is the period of
$\tan x$, we find $\alpha \approx -0.165 + \pi \approx 3.0$. Phase
shift is -3.0.

47. Since $a = -3$ and $b = -5$, the amplitude is
$\sqrt{(-3)^2 + (-5)^2} = \sqrt{34}$. If the terminal side
of α passes through $(-3, -5)$, then
$\tan \alpha = 5/3$. Using a calculator, one gets
$\tan^{-1}(5/3) \approx 1.03$ which is an angle in
quadrant I. Since $(-3, -5)$ is in quadrant III
and π is the period of $\tan x$, we obtain
$\alpha \approx 1.03 + \pi \approx 4.2$. The phase shift is -4.2.

49. By using a sum-to-product identity, we get

$$\frac{\sin(3t) - \sin(t)}{\cos(3t) + \cos(t)} =$$

$$\frac{2 \cos\left(\dfrac{3t + t}{2}\right) \sin\left(\dfrac{3t - t}{2}\right)}{2 \cos\left(\dfrac{3t + t}{2}\right) \cos\left(\dfrac{3t - t}{2}\right)} =$$

$$\frac{2 \cos(2t) \sin t}{2 \cos(2t) \cos t} =$$

$$\tan t$$

51. Using a sum-to-product identity, we find

$$\frac{\cos x - \cos(3x)}{\cos x + \cos(3x)} =$$

$$\frac{-2 \sin\left(\dfrac{x + 3x}{2}\right) \sin\left(\dfrac{x - 3x}{2}\right)}{2 \cos\left(\dfrac{x + 3x}{2}\right) \cos\left(\dfrac{x - 3x}{2}\right)} =$$

$$\frac{-2 \sin(2x) \sin(-x)}{2 \cos(2x) \cos(-x)} =$$

$$\frac{2 \sin(2x) \sin x}{2 \cos(2x) \cos x} =$$

$$\tan(2x) \tan(x)$$

53. By using a product-to-sum identity, we get

$$-\sin(x+y)\sin(x-y)$$

$$= -\frac{1}{2}\Big[\cos\left((x+y)-(x-y)\right) - \cos\left((x+y)+(x-y)\right)\Big]$$

$$= -\frac{1}{2}\Big[\cos\left(2y\right)-\cos\left(2x\right)\Big]$$

$$= -\frac{1}{2}\Big[(2\cos^2 y-1)-(2\cos^2 x-1)\Big]$$

$$= -\frac{1}{2}\Big[2\cos^2 y-2\cos^2 x\Big]$$

$$= \cos^2 x-\cos^2 y.$$

55. Let $A=\dfrac{x+y}{2}$ and $B=\dfrac{x-y}{2}$. Note,

$$A+B=x \quad \text{and} \quad A-B=y.$$

Expand the left-hand side and use product-to-sum identities.

$$(\sin A+\cos A)\,(\sin B+\cos B) =$$

$$\sin A\sin B+\sin A\cos B+$$

$$\cos A\sin B+\cos A\cos B =$$

$$\frac{1}{2}\Big[\cos(A-B)-\cos(A+B)\Big]+$$

$$\frac{1}{2}\Big[\sin(A+B)+\sin(A-B)\Big]+$$

$$\frac{1}{2}\Big[\sin(A+B)-\sin(A-B)\Big]+$$

$$\frac{1}{2}\Big[\cos(A-B)+\cos(A+B)\Big] =$$

$$\frac{1}{2}\Big[\cos y-\cos x\Big]+\frac{1}{2}\Big[\sin x+\sin y\Big]+$$

$$\frac{1}{2}\Big[\sin x-\sin y\Big]+\frac{1}{2}\Big[\cos y+\cos x\Big] =$$

$$\frac{1}{2}\Big[2\cos y+2\sin x\Big] =$$

$$\sin x+\cos y$$

57. Apply a sum-to-product identity in the 5th line.

$$\sin^2(A+B)-\sin^2(A-B)$$

$$= \sin(2A)\sin(2B) \quad \text{by Exercise 56}$$

$$= (2\sin A\cos A)(2\sin B\cos B)$$

$$= [2\cos A\cos B]\cdot[2\sin A\sin B]$$

$$= \Big[\cos(A-B)+\cos(A+B)\Big]\cdot$$

$$\Big[\cos(A-B)-\cos(A+B)\Big]$$

$$= \cos^2(A-B)-\cos^2(A+B)$$

59. Note that x can be written in the form $x=a\sin(t+\alpha)$. The maximum displacement of $x=\sqrt{3}\sin t+\cos t$ is

$$a=\sqrt{\sqrt{3}^2+1^2}=2.$$

Thus, 2 meters is the maximum distance between the block and its resting position.

Since the terminal side of α goes through $(\sqrt{3},1)$, we get $\tan\alpha=1/\sqrt{3}$ and one can choose $\alpha=\pi/6$. Then $x=2\sin(t+\pi/6)$.

61. Substitute $A=B=x$ in the identity

$$\cos A\cos B=\frac{1}{2}\left[\cos(A-B)+\cos(A+B)\right].$$

Then

$$\cos^2 x = \frac{1}{2}\left[1+\cos 2x\right]$$

$$\cos 2x = 2\cos^2 x-1$$

$$\cos 2x = 2\cos^2 x-(\cos^2 x+\sin^2 x)$$

$$\cos 2x = \cos^2 x-\sin^2 x$$

65. $\tan\dfrac{x}{2}=\dfrac{1-\cos x}{\sin x}=\dfrac{1-(-1/3)}{-2\sqrt{2}/3}=$

$$\dfrac{4/3}{-2\sqrt{2}/3}=-\dfrac{2}{\sqrt{2}}=-\sqrt{2}$$

67. a) $\sin(3.5+2.1)=\sin 5.6$

b) $\sin(2x-x)=\sin x$

c) $\sin(2\cdot 4.8)=\sin 9.6$

69. The acute angles are $\arcsin(12/13)\approx 67.4°$ and $\arcsin(5/13)\approx 22.6°$

71. a) Let $m = l + 1$ and $n = w + 1$ where l and w are relatively prime. The number of streets that the crow passes as it flies from $(1,1)$ to (m, n) is l. When the crow crosses an avenue, it will cross over two blocks bounded by the same streets (and not over an intersection) since l and w are relatively prime. Since the crow will cross over $w - 1$ avenues between the first block and the last block, the crow flies over $w - 1$ additional blocks. Thus, the total number of blocks that the crow flies over is

$$l + w - 1 = m + n - 3.$$

b) Let $m = l + 1$ and $n = w + 1$ where d is the greatest common divisor of l and w. Write $l = dl_1$ and $w = dw_1$ where l_1 and w_1 are relatively prime. Then we can break the flight of the crow into d segments described as follows:

1. $(1,1)$ to $(l_1 + 1, w_1 + 1)$
2. $(l_1 + 1, w_1 + 1)$ to $(2l_1 + 1, 2w_1 + 1)$, and so on, and finally from
3. $((d-1)l_1 + 1, (d-1)w_1 + 1)$ to $(dl_1 + 1, dw_1 + 1)$.

By part a), the number of blocks that the crow flies over in each segment is

$$l_1 + w_1 - 1.$$

Thus, the total number of blocks that the crow flies over is

$$d(l_1 + w_1 - 1) = l + w - d = m + n - 2 - d.$$

Chapter 3 Review Exercises

1. $1 - \sin^2 \alpha = \cos^2 \alpha$

3. $(1 - \csc x)(1 + \csc x) = 1 - \csc^2 x = -\cot^2 x$

5.
$$\frac{1}{1 + \sin \alpha} + \frac{\sin \alpha}{\cos^2 \alpha} = \frac{1}{1 + \sin \alpha} + \frac{\sin \alpha}{1 - \sin^2 \alpha} =$$
$$\frac{(1 - \sin \alpha) + \sin \alpha}{1 - \sin^2 \alpha} = \frac{1}{\cos^2 \alpha} = \sec^2 \alpha$$

7. $\tan(4s)$, by the double angle identity for tangent

9. $\sin(3\theta - 6\theta) = \sin(-3\theta) = -\sin(3\theta)$

11. $\tan\left(\dfrac{2z}{2}\right) = \tan z$, by a double-angle identity for tangent

13. Note, $\sin \alpha = \sqrt{1 - \left(\dfrac{-5}{13}\right)^2} = \sqrt{1 - \dfrac{25}{169}} =$

$\sqrt{\dfrac{144}{169}} = \dfrac{12}{13}$. Then $\tan \alpha = \dfrac{12/13}{-5/13} = -\dfrac{12}{5}$,

$\cot \alpha = -\dfrac{5}{12}$, $\csc \alpha = \dfrac{13}{12}$, $\sec \alpha = -\dfrac{13}{5}$.

15. By using a cofunction identity, we get

$\cos \alpha = \dfrac{-3}{5}$. Then $\sin \alpha = -\sqrt{1 - \left(\dfrac{-3}{5}\right)^2} =$

$-\sqrt{1 - \dfrac{9}{25}} = -\sqrt{\dfrac{16}{25}} = -\dfrac{4}{5}$, $\sec \alpha = -\dfrac{5}{3}$,

$\csc \alpha = -\dfrac{5}{4}$, $\tan \alpha = \dfrac{-4/5}{-3/5} = \dfrac{4}{3}$, $\cot \alpha = \dfrac{3}{4}$.

17. By the half-angle identity for sine, we find

$$\begin{aligned}
\sqrt{\frac{1 - \cos \alpha}{2}} &= \frac{3}{5} \\
\frac{1 - \cos \alpha}{2} &= \frac{9}{25} \\
1 - \cos \alpha &= \frac{18}{25} \\
\cos \alpha &= \frac{7}{25}.
\end{aligned}$$

Since $\dfrac{3\pi}{2} < \alpha < 2\pi$, α is in quadrant IV

and $\sin \alpha = -\sqrt{1 - \left(\dfrac{7}{25}\right)^2} = -\sqrt{1 - \dfrac{49}{625}} =$

$-\sqrt{\dfrac{576}{625}} = -\dfrac{24}{25}$. Then $\tan \alpha = \dfrac{-24/25}{7/25} =$

$-\dfrac{24}{7}$, $\cot \alpha = -\dfrac{7}{24}$, $\sec \alpha = \dfrac{25}{7}$, $\csc \alpha = -\dfrac{25}{24}$.

19. It is an identity.

$$\begin{aligned}
(\sin x + \cos x)^2 &= \\
\sin^2 x + 2 \sin x \cos x + \cos^2 x &= \\
1 + 2 \sin x \cos x &= \\
1 + \sin(2x)
\end{aligned}$$

21. It is not an identity since $\csc^2 x - \cot^2 x = 1$ and $\tan^2 - \sec^2 x = -1$.

23. Odd, since $f(-x) = \dfrac{\sin(-x) - \tan(-x)}{\cos(-x)} =$

$\dfrac{-\sin x + \tan x}{\cos x} = -\dfrac{\sin x - \tan x}{\cos x} = -f(x)$

25. It is neither even nor odd. Since $f(\pi/4) =$

$\dfrac{\cos(\pi/4) - \sin(\pi/4)}{\sec(\pi/4)} = \dfrac{\sqrt{2}/2 - \sqrt{2}/2}{\sqrt{2}} = 0$

and $f(-\pi/4) = \dfrac{\cos(-\pi/4) - \sin(-\pi/4)}{\sec(-\pi/4)} =$

$\dfrac{\sqrt{2}/2 + \sqrt{2}/2}{\sqrt{2}} = \dfrac{\sqrt{2}}{\sqrt{2}} = 1$, we find

$$f(\pi/4) \neq \pm f(-\pi/4).$$

27. Even, since $f(-x) = \dfrac{\sin(-x)\tan(-x)}{\cos(-x) + \sec(-x)} =$

$\dfrac{(-\sin x)(-\tan x)}{\cos x + \sec x} = \dfrac{\sin x \tan x}{\cos x + \sec x} = f(x)$

29.

$$\dfrac{1 + \tan^2 \theta}{1 - \tan^2 \theta}$$

$$= \dfrac{\dfrac{\sec^2 \theta}{1 - \dfrac{\sin^2 \theta}{\cos^2 \theta}} \cdot \dfrac{\cos^2 \theta}{\cos^2 \theta}}{}$$

$$= \dfrac{1}{\cos^2 \theta - \sin^2 \theta}$$

$$= \dfrac{1}{\cos 2\theta}$$

$$= \sec 2\theta$$

31.

$$\dfrac{\csc^2 x - \cot^2 x}{2\csc^2 x + 2\csc x \cot x}$$

$$= \dfrac{1}{\dfrac{2}{\sin^2 x} + 2 \cdot \dfrac{1}{\sin x} \cdot \dfrac{\cos x}{\sin x}}$$

$$= \dfrac{1}{\dfrac{2}{\sin^2 x} + \dfrac{2\cos x}{\sin^2 x}} \cdot \dfrac{\sin^2 x}{\sin^2 x}$$

$$= \dfrac{\sin^2 x}{2 + 2\cos x}$$

$$= \dfrac{1 - \cos^2 x}{2(1 + \cos x)}$$

$$= \dfrac{(1 - \cos x)(1 + \cos x)}{2(1 + \cos x)}$$

$$= \dfrac{1 - \cos x}{2}$$

$$= \sin^2 \left(\dfrac{x}{2}\right)$$

33.

$$\cot(\alpha - 45°) =$$

$$(\tan(\alpha - 45°))^{-1} =$$

$$\left(\dfrac{\tan \alpha - \tan 45°}{1 + \tan \alpha \tan 45°}\right)^{-1} =$$

$$\left(\dfrac{\tan \alpha - 1}{1 + \tan \alpha}\right)^{-1} =$$

$$\dfrac{1 + \tan \alpha}{\tan \alpha - 1}$$

35.

$$\dfrac{\sin 2\beta}{2\csc \beta} =$$

$$\dfrac{2\sin \beta \cos \beta}{2/\sin \beta} \cdot \dfrac{\sin \beta}{\sin \beta} =$$

$$\sin^2 \beta \cos \beta$$

37. Factor the numerator on the left-hand side as a difference of two cubes. Note, $\cot y \tan y = 1$.

$$\dfrac{\cot^3 y - \tan^3 y}{\sec^2 y + \cot^2 y} =$$

$$\dfrac{(\cot y - \tan y)(\cot^2 y + 1 + \tan^2 y)}{\sec^2 y + \cot^2 y} =$$

$$\dfrac{(\cot y - \tan y)(\cot^2 y + \sec^2 y)}{\sec^2 y + \cot^2 y} =$$

$$\cot y - \tan y =$$

$$\dfrac{1}{\tan y} - \tan y =$$

$$\dfrac{1 - \tan^2 y}{\tan y} =$$

$$2 \cdot \dfrac{1 - \tan^2 y}{2\tan y} =$$

$$2 \cdot (\tan 2y)^{-1} \;=\;$$
$$2 \cot(2y)$$

39. Rewrite the left side.

$$\frac{-1}{\sin x} + \frac{1}{\cos x} \;=\;$$

$$\frac{-\cos x + \sin x}{\sin x \cos x} \;=\;$$

$$\frac{-2\cos x + 2\sin x}{2\sin x \cos x} \;=\;$$

$$\frac{2\sin x - 2\cos x}{\sin 2x}$$

41. Rewrite the left side.

$$\frac{\cos x - \sin x}{\cos x + \sin x} \;=\;$$

$$\frac{(\cos x - \sin x)^2}{\cos^2 x - \sin^2 x} \;=\;$$

$$\frac{1 - \sin 2x}{\cos 2x} \;=\;$$

$$\frac{1 - \sin^2 2x}{\cos 2x(1 + \sin 2x)} \;=\;$$

$$\frac{\cos^2 2x}{\cos 2x(1 + \sin 2x)} \;=\;$$

$$\frac{\cos 2x}{1 + \sin 2x}$$

43. By using double-angle identities, we obtain

$$\cos(2 \cdot 2x) \;=\;$$
$$1 - 2\sin^2(2x) \;=\;$$
$$1 - 2\left(2\sin x \cos x\right)^2 \;=\;$$
$$1 - 8\sin^2 x \cos^2 x \;=\;$$
$$1 - 8\sin^2 x(1 - \sin^2 x) \;=\;$$
$$8\sin^4 x - 8\sin^2 x + 1$$

45. By the double-angle identity for sine, we get

$$\sin^4(2x) \;=\;$$
$$(2\sin x \cos x)^4 \;=\;$$
$$16\sin^4 x \cos^4 x \;=\;$$
$$16\sin^4 x(1 - \sin^2 x)^2 \;=\;$$
$$16\sin^4 x(1 - 2\sin^2 x + \sin^4 x) \;=\;$$
$$16\sin^4 x - 32\sin^6 x + 16\sin^8 x$$

47. $\tan\left(\dfrac{-\pi/6}{2}\right) = \dfrac{1 - \cos(-\pi/6)}{\sin(-\pi/6)} =$

$$\frac{1 - \sqrt{3}/2}{-1/2} \cdot \frac{2}{2} = \frac{2 - \sqrt{3}}{-1} = \sqrt{3} - 2$$

49. $\sin\left(\dfrac{-150°}{2}\right) = -\sqrt{\dfrac{1 - \cos(-150°)}{2}} =$

$$-\sqrt{\frac{1 - (-\sqrt{3}/2)}{2} \cdot \frac{2}{2}} = -\sqrt{\frac{2 + \sqrt{3}}{4}} =$$

$$-\frac{\sqrt{2 + \sqrt{3}}}{2} \text{ or alternatively by using a}$$
difference identity we find

$$\sin\left(-30° - 45°\right) = -\frac{\sqrt{2} + \sqrt{6}}{4}$$

51. Let $a = 4$, $b = 4$, and $r = \sqrt{4^2 + 4^2} = 4\sqrt{2}$. If the terminal side of α goes through $(4, 4)$, then $\tan \alpha = 4/4 = 1$ and one can choose $\alpha = \pi/4$. So

$$y = 4\sqrt{2}\sin(x + \pi/4),$$

amplitude is $4\sqrt{2}$, period is 2π, and phase shift is $-\pi/4$.

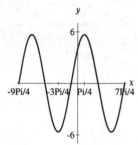

53. Let $a = -2$, $b = 1$, and $r = \sqrt{(-2)^2 + 1^2} = \sqrt{5}$. If the terminal side of α goes through $(-2, 1)$, then $\tan \alpha = -1/2$. Since $\tan^{-1}(-1/2) \approx -0.46$ and $(-2, 1)$ is in quadrant II, one can choose $\alpha = \pi - 0.46 = 2.68$. So $y = \sqrt{5}\sin(x + 2.68)$, amplitude is $\sqrt{5}$, period is 2π, and phase shift is -2.68.

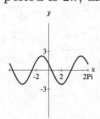

55. $\cos 15° + \cos 19° =$

$$2 \cos \left(\frac{15° + 19°}{2} \right) \cos \left(\frac{15° - 19°}{2} \right) =$$

$$2 \cos 17° \cos(-2°) = 2 \cos 17° \cos 2°$$

57. $\sin(\pi/4) - \sin(-\pi/8) = \sin(\pi/4) + \sin(\pi/8) =$

$$2 \sin \left(\frac{\pi/4 + \pi/8}{2} \right) \cos \left(\frac{\pi/4 - \pi/8}{2} \right) =$$

$$2 \sin \left(\frac{3\pi/8}{2} \right) \cos \left(\frac{\pi/8}{2} \right) =$$

$$2 \sin \left(\frac{3\pi}{16} \right) \cos \left(\frac{\pi}{16} \right)$$

59. $2 \sin 11° \cos 13° =$

$$\sin(11° + 13°) + \sin(11° - 13°) =$$

$$\sin 24° + \sin(-2°) = \sin 24° - \sin 2°$$

61. $2 \cos \dfrac{x}{4} \cos \dfrac{x}{3} =$

$$\cos \left(\frac{x}{4} - \frac{x}{3} \right) + \cos \left(\frac{x}{4} + \frac{x}{3} \right) =$$

$$\cos \left(-\frac{x}{12} \right) + \cos \left(\frac{7x}{12} \right) =$$

$$\cos \left(\frac{x}{12} \right) + \cos \left(\frac{7x}{12} \right)$$

63. Note, y can be written in the form

$y = a \sin(t + \alpha)$ where $a = \sqrt{2^2 + 1^2} = \sqrt{5}$

and $\alpha = \cos^{-1} \left(\dfrac{2}{\sqrt{5}} \right) \approx 0.46$.

Thus, $y = \sqrt{5} \sin(t + 0.46)$ and the maximum height above its normal position is $\sqrt{5}$ in.

65. Consider the list of 6 digit numbers from 000,000 through 999,999. There are 1 million 6 digit numbers in this list for a total of 6 million digits. Each of the ten digits 0 through 9 occurs with the same frequency in this list. So there are 600,000 of each in this list. In particular there are 600,000 ones in the list. You need one more to write 1,000,000. So there are 600,001 ones used in writing the numbers 1 through 1 million.

Chapter 3 Test

1. $\dfrac{1}{\cos x} \cdot \dfrac{\cos x}{\sin x} \cdot 2 \sin x \cos x = 2 \cos x$

2. $\sin(2t + 5t) = \sin(7t)$

3. $\dfrac{1}{1 - \cos y} + \dfrac{1}{1 + \cos y} =$

$$\frac{1 + \cos y + 1 - \cos y}{1 - \cos^2 y} =$$

$$\frac{2}{\sin^2 y} = 2 \csc^2 y$$

4. $\tan \left(\dfrac{\pi}{5} + \dfrac{\pi}{10} \right) = \tan \left(\dfrac{3\pi}{10} \right)$

5.

$$\frac{\sin \beta \cos \beta}{\sin \beta / \cos \beta} =$$

$$\sin \beta \cos \beta \cdot \frac{\cos \beta}{\sin \beta} =$$

$$\cos^2 \beta =$$

$$1 - \sin^2 \beta$$

6.

$$\frac{1}{\sec \theta - 1} - \frac{1}{\sec \theta + 1} =$$

$$\frac{\sec \theta + 1 - (\sec \theta - 1)}{\sec^2 \theta - 1} =$$

$$\frac{2}{\tan^2 \theta} =$$

$$2 \cot^2 \theta$$

7. Using the cofunction identity for cosine, we obtain

$$\cos(\pi/2 - x) \cos(-x) =$$

$$\sin x \cos x =$$

$$\frac{2 \sin x \cos x}{2} =$$

$$\frac{\sin(2x)}{2}$$

8. Factor the left-hand side and use a half-angle identity for tangent.

$$\tan(t/2) \cdot (\cos^2 t - 1) \ =$$
$$\frac{1 - \cos t}{\sin t} \cdot (-\sin^2 t) \ =$$
$$(1 - \cos t) \cdot (-\sin t) \ =$$
$$(\cos t - 1)\sin t \ =$$
$$\cos t \sin t - \sin t \ =$$
$$\frac{\sin t}{\sec t} - \sin t$$

9. Let $a = 1$, $b = -\sqrt{3}$, $r = \sqrt{1^2 + (-\sqrt{3})^2} = 2$. If the terminal side of α goes through $(1, -\sqrt{3})$, then $\tan \alpha = -\sqrt{3}$ and one can choose $\alpha = 5\pi/3$. Then

$$y = 2\sin(x + 5\pi/3),$$

the period is 2π, amplitude is 2, and the phase shift is $-5\pi/3$.

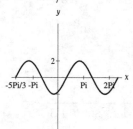

10. If $\csc \alpha = 2$, then $\sin \alpha = \dfrac{1}{2}$.

Since α is in quadrant II, we obtain

$$\cos \alpha = -\sqrt{1 - \left(\frac{1}{2}\right)^2} = -\sqrt{1 - \frac{1}{4}} =$$
$$-\sqrt{\frac{3}{4}} = -\frac{\sqrt{3}}{2}, \ \sec \alpha = -\frac{2}{\sqrt{3}} = -\frac{2\sqrt{3}}{3},$$
$$\tan \alpha = \frac{1/2}{-\sqrt{3}/2} = -\frac{1}{\sqrt{3}} = -\frac{\sqrt{3}}{3},$$
and $\cot \alpha = -\sqrt{3}$.

11. Even, $f(-x) = (-x)\sin(-x) = (-x)(-\sin x) = x\sin x = f(x)$.

12. Using a half-angle identity, we obtain

$$\sin\left(\frac{-\pi/6}{2}\right) = -\sqrt{\frac{1 - \cos(-\pi/6)}{2}} =$$

$$-\sqrt{\frac{1 - \sqrt{3}/2}{2} \cdot \frac{2}{2}} = -\sqrt{\frac{2 - \sqrt{3}}{4}} =$$

$$-\frac{\sqrt{2 - \sqrt{3}}}{2}.$$ Equivalently, using a difference identity we find

$$\sin\left(-\frac{\pi}{12}\right) = \sin\left(\frac{\pi}{4} - \frac{\pi}{3}\right) = \frac{\sqrt{2} - \sqrt{6}}{4}.$$

13. If $x = y = \pi/6$, then we find that

$$\tan x + \tan y = 2\tan(\pi/6) = 2 \cdot \frac{\sqrt{3}}{3} \text{ and}$$

$$\tan(x+y) = \tan(\pi/6 + \pi/6) = \tan(\pi/3) = \sqrt{3}.$$
Thus, it is not an identity.

Tying It All Together

1. $\dfrac{\sqrt{2}}{2}$ **2.** $\dfrac{\sqrt{2}}{2}$

3. $\dfrac{1}{2}$ **4.** $\dfrac{1}{2}$

5. $\dfrac{1}{2}$ **6.** $\dfrac{1}{2}$

7. $\sin\left(\dfrac{55\pi}{6}\right) = \sin\left(\dfrac{7\pi}{6}\right) = -\dfrac{1}{2}$

8. $\sin\left(-\dfrac{23\pi}{2}\right) = \sin\left(\dfrac{\pi}{2}\right) = 1$

9. -1 **10.** 0

11. Odd, $f(-x) = (-x)^3 + \sin(-x) = -x^3 - \sin x = -f(x)$

12. Even, $f(-x) = (-x)^3 \sin(-x) = (-x^3)(-\sin x) = x^3 \sin x = f(x)$

13. Even, $f(-x) = \dfrac{\sin(-x)}{-x} = \dfrac{-\sin x}{-x} = \dfrac{\sin x}{x} = f(x)$

14. Even, $f(-x) = |\sin(-x)| = |-\sin x| = |\sin x| = f(x)$

15. Even, since $y = \cos^5 x$, $y = \cos^3 x$, $y = 2\cos x$ are even functions, and a sum and difference of even functions is again an even function.

16. Odd, since

$$\begin{aligned} f(-x) &= (-x)^3 \sin^4(-x) + (-x)\sin^2(-x) \\ &= -x^3 \sin^4(x) - x\sin^2(x) \\ f(-x) &= -f(x) \end{aligned}$$

17. Let $\alpha = \beta = \dfrac{\pi}{2}$. Then $\sin(\alpha + \beta) = \sin \pi = 0$ while $\sin \alpha + \sin \beta = 1 + 1 = 2$. Thus, $\sin(\alpha + \beta) = \sin \alpha + \sin \beta$ is not an identity.

18. Let $\alpha = \dfrac{\pi}{2}$ and $\beta = \dfrac{\pi}{4}$. Then

$$\cos(\alpha - \beta) = \cos\left(\frac{\pi}{2} - \frac{\pi}{4}\right) = \cos\frac{\pi}{4} = \frac{\sqrt{2}}{2}$$

while $\cos\alpha - \cos\beta = 0 - \dfrac{\sqrt{2}}{2} = -\dfrac{\sqrt{2}}{2}$. Thus, $\cos(\alpha - \beta) = \cos\alpha - \cos\beta$ is not an identity.

19. Note $\sin^{-1}(0.1) \approx 0.1$ and $\dfrac{1}{\sin 0.1} \approx 10.0$. Thus, $\sin^{-1} x = \dfrac{1}{\sin x}$ is not an identity.

20. Note $\sin^2(2) > 0$ and $\sin(2^2) = \sin(4) < 0$. Thus, $\sin^2 x = \sin(x^2)$ is not an identity.

21. Form a right triangle with $\alpha = 30°$, $a = 4$.

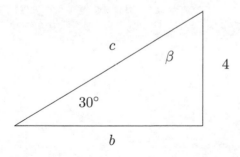

Since $\tan 30° = 4/b$ and $\sin 30° = 4/c$, we get

$$b = \frac{4}{\tan 30°} = 4\sqrt{3}$$

and

$$c = \frac{4}{\sin 30°} = 8.$$

Also, $\beta = 90° - 30° = 60°$.

22. Form a right triangle with $a = \sqrt{3}$, $b = 1$.

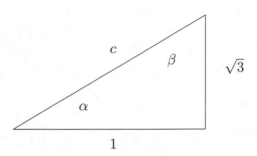

Then $c = \sqrt{\sqrt{3}^2 + 1^2} = 2$ by the Pythagorean Theorem. Since $\tan\alpha = \sqrt{3}/1 = \sqrt{3}$, we get $\alpha = 60°$ and $\beta = 30°$.

23. Form a right triangle with $b = 5$ as shown below.

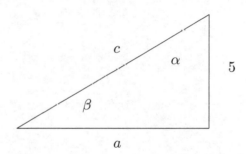

Since $\cos\beta = 0.3$, we find $\beta = \cos^{-1}(0.3) \approx 72.5°$ and $\alpha = 17.5°$. Since $\sin 72.5° = 5/c$ and $\tan 72.5° = 5/a$, we obtain $c = 5/\sin 72.5° \approx 5.2$ and $a = 5/\tan 72.5° \approx 1.6$.

24. Form a right triangle with $a = 2$.

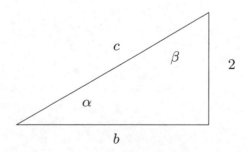

Since $\sin\alpha = 0.6$, we obtain $\alpha = \sin^{-1}(0.6) \approx 36.9°$ and $\beta = 53.1°$. Since $\sin 36.9° = 2/c$ and $\tan 36.9° = 2/b$, we get $c = 2/\sin 36.9° \approx 3.3$ and $b = 2/\tan 36.9° \approx 2.7$.

25. angle

26. central

27. acute

28. quadrantal

29. coterminal

30. minutes

31. seconds

32. unit

33. π

34. αr

For Thought

1. True, $\sin^{-1}(0) = 0 = \sin(0)$.

2. True, since $\sin(3\pi/4) = \dfrac{\sqrt{2}}{2} = \dfrac{1}{\sqrt{2}}$.

3. False, $\cos^{-1}(0) = \pi/2$.

4. False, $\sin^{-1}(\sqrt{2}/2) = 45°$.

5. False, since it equals $\tan^{-1}(1/5)$.

6. True, since $1/5 = 0.2$.

7. True, $\sin(\cos^{-1}(\sqrt{2}/2)) = \sin(\pi/4) = 1/\sqrt{2}$.

8. True by definition of $y = \sec^{-1}(x)$.

9. False, since $f^{-1}(x) = \sin(x)$ where $-\pi/2 \le x \le \pi/2$.

10. False, since the secant and cotangent functions are not invertible functions.

4.1 Exercises

1. domain

3. domain

5. $\sin^{-1}(y)$

7. $\pi/4$ **9.** $\pi/3$

11. $-\pi/4$ **13.** 0

15. $-45°$ **17.** $30°$

19. $0°$ **21.** $-90°$

23. $33.8°$

25. $-19.5°$

27. $0.6°$ **29.** $\pi/4$

31. $\pi/6$ **33.** $3\pi/4$

35. $\pi/2$ **37.** $86.6°$

39. $101.5°$ **41.** $89.4°$ **43.** $-\pi/4$

45. $\pi/6$ **47.** $\pi/3$ **49.** 1.571

51. $\pi/3$ **53.** $\pi/3$

55. $\dfrac{\pi}{3}$ **57.** $\dfrac{5\pi}{6}$

59. 0 **61.** $\dfrac{\pi}{2}$

63. $\dfrac{3\pi}{4}$ **65.** 1.87

67. 0.15 **69.** 2.72

71. 3.06 **73.** 2.36

75. $\tan(\pi/3) = \sqrt{3}$

77. $\sin^{-1}(-1/2) = -\pi/6$

79. $\cot^{-1}(\sqrt{3}) = \pi/6$

81. $\arcsin(\sqrt{2}/2) = \pi/4$

83. $\tan(\pi/4) = 1$

85. $\cos^{-1}(0) = \pi/2$

87. $\sin^{-1}(\sin 3\pi/4) = \sin^{-1}(\sqrt{2}/2) = \pi/4$

89. $\cos(\cos^{-1}(-\sqrt{3}/2)) = \cos(5\pi/6) = -\sqrt{3}/2$

91. $\tan^{-1}(\tan \pi) = \tan^{-1}(0) = 0$

93. 0.8930

95. Undefined

97. -0.9802

99. -0.4082

101. 3.4583

103. 1.0183

105. Let $\theta = \arccos x$. Then $\cos \theta = x$ and θ lies in quadrant 1 or 2. Since $\sin^2 \theta = 1 - \cos^2 \theta = 1 - x^2$, we obtain $\sin(\arccos x) = \sin \theta = \pm\sqrt{1 - x^2}$. Since sine is positive in both quadrants 1 and 2, we have $\sin(\arccos x) = \sqrt{1 - x^2}$.

107. Note, $\cos(\theta) = \dfrac{1}{\sec \theta} = \dfrac{1}{\pm\sqrt{\tan^2 \theta + 1}}$.

Since $\theta = \arctan x$ is an angle in quadrant 1 or 4, and cosine is positive in both quadrants 1 and 4, we get

$$\cos(\arctan x) = \dfrac{1}{\sqrt{\tan^2(\arctan x) + 1}} = \dfrac{1}{\sqrt{x^2 + 1}}.$$

109. Let $\theta = \text{arccot } x$ be an angle that lies in quadrant 1 or 2. Since cosine is positive in quadrants 1 and 2, we obtain

$$
\begin{aligned}
\cos(\text{arccot } x) &= \cos(\theta) \\
&= \sqrt{1 - \sin^2 \theta} \\
&= \sqrt{1 - \frac{1}{\csc^2 \theta}} \\
&= \sqrt{1 - \frac{1}{1 + \cot^2 \theta}} \\
&= \sqrt{1 - \frac{1}{1 + x^2}} \\
&= \sqrt{\frac{x^2}{1 + x^2}} \\
&= \frac{\pm x}{\sqrt{1 + x^2}}
\end{aligned}
$$

Note, $\cos(\text{arccot } x)$ is positive exactly when $x > 0$, and $\cos(\text{arccot } x)$ is negative exactly when $x < 0$. Thus,

$$\cos(\text{arccot } x) = \frac{x}{\sqrt{1 + x^2}}.$$

111. Note, $\tan x = \pm\sqrt{\sec^2 x - 1}$ and $\cos(\arcsin x) = \sqrt{1 - x^2}$. Then

$$
\begin{aligned}
\tan(\arcsin x) &= \pm\sqrt{\sec^2(\arcsin x) - 1} \\
&= \pm\sqrt{\left(\frac{1}{\sqrt{1 - x^2}}\right)^2 - 1} \\
&= \pm\sqrt{\frac{1}{1 - x^2} - 1} \\
&= \pm\sqrt{\frac{x^2}{1 - x^2}} \\
&= \pm\frac{\sqrt{x^2}}{\sqrt{1 - x^2}} \\
&= \pm\frac{\pm x}{\sqrt{1 - x^2}} \\
&= \pm\frac{x}{\sqrt{1 - x^2}}.
\end{aligned}
$$

Note, $\tan(\arcsin x)$ is positive exactly when $x > 0$, and $\tan(\arcsin x)$ is negative exactly when $x < 0$. Thus,

$$\tan(\arcsin x) = \frac{x}{\sqrt{1 - x^2}}.$$

113. Note, $\arctan x$ is an angle in quadrant 1 or 4, and secant is positive in quadrants 1 and 4. Since $\sec(\theta) = \pm\sqrt{\tan^2(\theta) + 1}$, we have

$$
\begin{aligned}
\sec(\arctan x) &= \sqrt{\tan^2(\arctan x) + 1} \\
&= \sqrt{x^2 + 1}.
\end{aligned}
$$

115. Consider the right triangle with hypotenuse 2400, altitude 2000, and the angle between the hypotenuse and the altitude is $\theta/2$. Since

$$\cos\left(\frac{\theta}{2}\right) = \frac{2000}{2400},$$

we find

$$
\begin{aligned}
\theta &= 2\cos^{-1}\left(\frac{2000}{2400}\right) \\
&\approx 67.1°
\end{aligned}
$$

Thus, the angle for which the airplane is within range of the gun is $\theta \approx 67.1°$.

117. If $t = 0.25$, then

$$u = 1 - 8(0.25) + 8(0.25)^2 = -0.5$$

Consequently,

$$
\begin{aligned}
P &= \frac{\cos^{-1}(-0.5) - \sin(\cos^{-1}(-0.5))}{\pi} \\
&= \frac{2\pi/3 - \sqrt{3}/2}{\pi} \\
&\approx 39\%
\end{aligned}
$$

119. Note, the domain of $y = \sin\left(\sin^{-1} x\right)$ is $[-1, 1]$ and

$$y = \sin\left(\sin^{-1} x\right) = x.$$

Then the graph of $y = \sin\left(\sin^{-1} x\right)$ is a segment of the line $y = x$ with endpoints $(-1, -1)$ and $(1, 1)$.

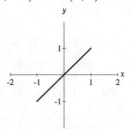

121. Since $\sin^{-1}(1/x) = \csc^{-1}(x)$ for $|x| \geq 1$, the graph of $y = \sin^{-1}(1/x)$ is the same as the graph of $y = \csc^{-1}(x)$.

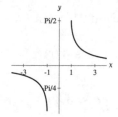

123. The angle opposite the shortest side is $26°$. If h is the hypotenuse, then

$$h = \frac{7}{\sin 26°} \approx 16.0.$$

If x is the side adjacent to $26°$, then

$$x = \frac{7}{\tan 26°} \approx 14.4$$

125. The amplitude is $A = 5$. Since

$$4x - \pi = 4\left(x - \frac{\pi}{4}\right)$$

the period is $\dfrac{2\pi}{B} = \dfrac{2\pi}{4} = \dfrac{\pi}{2}$ and the phase shift is $\dfrac{\pi}{4}$. The range is $[-5 - 3, 5 - 3]$ or $[-8, 2]$.

127. $3 - 3\sin^2 x = 3(1 - \sin^2 x) = 3\cos^2 x$

129. Note, if $(x, x + h)$ is such an ordered pair then the average is $x + h/2$. Since the average is not a whole number, then $h = 1$. Thus, the ordered pairs are $(4, 5)$, $(49, 50)$, $(499, 500)$, and $(4999, 5000)$.

For Thought

1. False, since $\sin \pi = 0$.   **2.**   True

3. False, since $\sin(\pi/3) = \dfrac{\sqrt{3}}{2}$.   **4.**   True

5. True   **6.**   True

7. False, only solutions are $45°$ and $315°$.

8. False, there is no solution in $[0, \pi)$.

9. True, since $-29°$ and $331°$ are coterminal angles.

10. True, since $\tan(3\pi/4) = \tan(7\pi/4) = -1$

and $\dfrac{3\pi}{4} = \dfrac{7\pi}{4} + (-1) \cdot \pi$.

4.2 Exercises

1. identity

3. $\left\{x \mid x = \dfrac{\pi}{2} + k\pi, k \text{ an integer}\right\}$

5. $\{x \mid x = 2k\pi, k \text{ an integer}\}$

7. No solution, since $\sin x = -2$ and the range of $y = \sin x$ is $[-1, 1]$.

9. $\left\{x \mid x = \dfrac{3\pi}{2} + 2k\pi, k \text{ an integer}\right\}$

11. $\{x \mid x = k\pi, k \text{ an integer}\}$

13. Solution in $[0, \pi)$ is $x = \dfrac{3\pi}{4}$. The solution set is $\left\{x \mid x = \dfrac{3\pi}{4} + k\pi\right\}$.

15. Solutions in $[0, 2\pi)$ are $x = \dfrac{\pi}{3}, \dfrac{5\pi}{3}$. So solution set is $\left\{x \mid x = \dfrac{\pi}{3} + 2k\pi \text{ or } x = \dfrac{5\pi}{3} + 2k\pi\right\}$.

17. Solutions in $[0, 2\pi)$ are $x = \dfrac{\pi}{4}, \dfrac{3\pi}{4}$. So solution set is $\left\{x \mid x = \dfrac{\pi}{4} + 2k\pi \text{ or } x = \dfrac{3\pi}{4} + 2k\pi\right\}$.

19. Since $\tan x = \dfrac{1}{\sqrt{3}} = \dfrac{\sqrt{3}}{3}$, the solution in $[0, \pi)$ is $x = \dfrac{\pi}{6}$. Thus, the solution set is $\left\{x \mid x = \dfrac{\pi}{6} + k\pi\right\}$.

21. Solutions in $[0, 2\pi)$ are $x = \dfrac{5\pi}{6}, \dfrac{7\pi}{6}$. Then the solution set is $\left\{x \mid x = \dfrac{5\pi}{6} + 2k\pi \text{ or } x = \dfrac{7\pi}{6} + 2k\pi\right\}$.

23. Since $\sin x = -\dfrac{\sqrt{2}}{2}$, the solutions in $[0, 2\pi)$

are $x = \dfrac{5\pi}{4}, \dfrac{7\pi}{4}$. Thus, the solution set is

$\left\{ x \mid x = \dfrac{5\pi}{4} + 2k\pi \text{ or } x = \dfrac{7\pi}{4} + 2k\pi \right\}$.

25. Since $\tan x = \sqrt{3}$, the solution in $[0, \pi)$

is $x = \dfrac{\pi}{3}$. Thus, the solution set is

$\left\{ x \mid x = \dfrac{\pi}{3} + k\pi \right\}$.

27. Solutions in $[0, 360°)$ are $\alpha = 90°, 270°$.
So solution set is $\{\alpha \mid \alpha = 90° + k \cdot 180°\}$.

29. Solution in $[0, 360°)$ is $\alpha = 90°$. So the
solution set is $\{\alpha \mid \alpha = 90° + k \cdot 360°\}$.

31. Solution in $[0, 180°)$ is $\alpha = 0°$.
The solution set is $\{\alpha \mid \alpha = k \cdot 180°\}$.

33. Since $\cos \alpha = \dfrac{\sqrt{2}}{2}$, the solutions in $[0, 360°)$ are

$\alpha = 45°, 315°$. The solution set is
$\{\alpha \mid \alpha = 45° + k \cdot 360°, \alpha = 315° + k \cdot 360°\}$.

35. Since $\sin \alpha = -\dfrac{1}{2}$, the solutions in $[0, 360°)$ are

$\alpha = 210°, 330°$. The solution set is
$\{\alpha \mid \alpha = 210° + k \cdot 360°, \alpha = 330° + k \cdot 360°\}$.

37. Since $\tan \alpha = 1$, the solution in $[0, 180°)$
is $\alpha = 45°$. The solution set is
$\{\alpha \mid \alpha = 45° + k \cdot 180°\}$.

39. One solution is $\cos^{-1}(0.873) \approx 29.2°$. Another
solution is $360° - 29.2° = 330.8°$. The solution
set is $\{29.2°, 330.8°\}$.

41. One solution is $\sin^{-1}(-0.244) \approx -14.1°$.
This is coterminal with $345.9°$. Another
solution is $180° + 14.1° = 194.1°$. The solution
set is $\{345.9°, 194.1°\}$.

43. One solution is $\tan^{-1}(5.42) \approx 79.5°$.
Another solution is $180° + 79.5° = 259.5°$.
The solution set is $\{79.5°, 259.5°\}$.

45. No solution, since $\cos \alpha = \sqrt{3} > 1$ and the
range of $y = \cos \alpha$ is $[-1, 1]$.

47. Since $x = \cos^{-1}(0.66) \approx 0.85$ and
$2\pi - \cos^{-1}(0.66) \approx 5.43$, the solution set
is $\{0.85, 5.43\}$.

49. If $a = \sin^{-1}(-1/4) \approx -0.25$, then

$$x = 2\pi + a \approx 6.03$$

and

$$x = \pi - a \approx 3.39$$

are solutions to $\sin x = -1/4$.
The solution set $\{3.39, 6.03\}$.

51. Since $\tan^{-1}(1/\sqrt{6}) \approx 0.39$ and
$\pi + \tan^{-1}(1/\sqrt{6}) \approx 3.53$, the
solution set is $\{0.39, 3.53\}$.

53. Let $a = \cos^{-1}(-2/\sqrt{5}) \approx 2.68$.
Then $x = a$ and $x = 2\pi - a \approx 3.61$
are solutions to $\cos x = -2/\sqrt{5}$.
The solution set is $\{2.68, 3.61\}$.

55. Since $\cos \alpha = 1$, the solution
set is $\{-360°, 0°, 360°\}$.

57. Since $\sin \beta = -1/2$, the solution
set is $\{210°, 330°\}$.

59. Multiplying by the LCD, we get

$$
\begin{aligned}
25.9 \sin \alpha &= 23.4 \sin 67.2° \\
\sin \alpha &= \frac{23.4 \sin 67.2°}{25.9} \\
\alpha &= \sin^{-1}\left(\frac{23.4 \sin 67.2°}{25.9} \right) \\
\alpha &\approx 56.4°.
\end{aligned}
$$

The solution set is $\{56.4°\}$.

61. Isolate $\cos \alpha$ on one side.

$$
\begin{aligned}
2(5.4)(8.2) \cos \alpha &= 5.4^2 + 8.2^2 - 3.6^2 \\
\cos \alpha &= \frac{5.4^2 + 8.2^2 - 3.6^2}{2(5.4)(8.2)}
\end{aligned}
$$

Then

$$
\begin{aligned}
\alpha &= \cos^{-1}\left(\frac{5.4^2 + 8.2^2 - 3.6^2}{2(5.4)(8.2)} \right) \\
\alpha &\approx 19.6°.
\end{aligned}
$$

The solution set is $\{19.6°\}$.

63. Since $\cos 3y = x/2$, we obtain

$$3y = \cos^{-1}(x/2)$$
$$y = \frac{1}{3}\cos^{-1}(x/2).$$

65. Since

$$\sin m = \frac{t-2}{-6} = \frac{2-t}{6}$$

we have

$$m = \sin^{-1}\left(\frac{2-t}{6}\right).$$

67. Since

$$\tan(a/3) = \frac{b+d}{7}$$

we find

$$a/3 = \tan^{-1}\left(\frac{b+d}{7}\right)$$
$$a = 3\tan^{-1}\left(\frac{b+d}{7}\right).$$

69. Since

$$\sin(\pi b - \pi) = \frac{q}{3}$$

we get

$$\pi b - \pi = \sin^{-1}\left(\frac{q}{3}\right)$$

$$\pi b = \sin^{-1}\left(\frac{q}{3}\right) + \pi$$

$$b = \frac{1}{\pi}\sin^{-1}\left(\frac{q}{3}\right) + 1.$$

71. To find f^{-1}, interchange x and y, solve for y, and replace y by $f^{-1}(x)$:

$$x = \sin 2y$$
$$\sin^{-1}x = 2y$$
$$\frac{1}{2}\sin^{-1}x = y$$
$$f^{-1}(x) = \frac{1}{2}\sin^{-1}x.$$

Since the domain of f is

$$-\frac{\pi}{4} \le x \le \frac{\pi}{4}$$

we obtain

$$-\frac{\pi}{2} \le 2x \le \frac{\pi}{2}$$
$$-1 \le \sin 2x \le 1.$$

Then the range of f is $[-1,1]$. Thus, the domain and range of f^{-1} are $[-1,1]$ and $[-\pi/4, \pi/4]$, respectively.

73. To find f^{-1}, interchange x and y, solve for y, and replace y by $f^{-1}(x)$:

$$x = 2\cos 3y$$
$$\cos^{-1}(x/2) = 3y$$
$$\frac{1}{3}\cos^{-1}(x/2) = y$$
$$f^{-1}(x) = \frac{1}{3}\cos^{-1}(x/2).$$

Since the domain of f is

$$0 \le x \le \frac{\pi}{3}$$

we find

$$0 \le 3x \le \pi$$
$$-1 \le \cos 3x \le 1$$
$$-2 \le 2\cos 3x \le 2.$$

Then the range of f is $[-2,2]$. Thus, the domain and range of f^{-1} are $[-2,2]$ and $[0, \pi/3]$, respectively.

75. To find f^{-1}, interchange x and y, solve for y, and replace y by $f^{-1}(x)$:

$$x = 3 + \tan(\pi y)$$
$$\tan^{-1}(x-3) = \pi y$$
$$\frac{1}{\pi}\tan^{-1}(x-3) = y$$
$$f^{-1}(x) = \frac{1}{\pi}\tan^{-1}(x-3).$$

Since the domain of f is

$$-\frac{1}{2} < x < \frac{1}{2}$$

we find

$$-\frac{\pi}{2} < \pi x < \frac{\pi}{2}$$
$$-\infty < \tan \pi x < \infty$$
$$-\infty < 3 + \tan \pi x < \infty.$$

Then the range of f is $(-\infty, \infty)$. Thus, the domain and range of f^{-1} are $(-\infty, \infty)$ and $(-1/2, 1/2)$, respectively.

77. To find f^{-1}, interchange x and y, solve for y, and replace y by $f^{-1}(x)$:

$$
\begin{aligned}
x &= 2 - \sin(\pi y - \pi) \\
\pi y - \pi &= \sin^{-1}(2 - x) \\
\pi y &= \sin^{-1}(2 - x) + \pi \\
y &= \frac{1}{\pi}\sin^{-1}(2 - x) + 1 \\
f^{-1}(x) &= \frac{1}{\pi}\sin^{-1}(2 - x) + 1.
\end{aligned}
$$

Since the domain of f is

$$\frac{1}{2} \leq x \leq \frac{3}{2}$$

we obtain

$$
\begin{aligned}
\frac{\pi}{2} \leq \pi x &\leq \frac{3\pi}{2} \\
-\frac{\pi}{2} \leq \pi x - \pi &\leq \frac{\pi}{2} \\
-1 \leq \sin(\pi x - \pi) &\leq 1 \\
-1 \leq -\sin(\pi x - \pi) &\leq 1 \\
1 \leq 2 - \sin(\pi x - \pi) &\leq 3.
\end{aligned}
$$

Then the range of f is $[1, 3]$. Thus, the domain and range of f^{-1} are $[1, 3]$ and $[1/2, 3/2]$, respectively.

79. To find f^{-1}, interchange x and y, solve for y, and replace y by $f^{-1}(x)$:

$$
\begin{aligned}
x &= \sin^{-1}(3y) \\
\sin x &= 3y \\
\frac{1}{3}\sin x &= y \\
f^{-1}(x) &= \frac{1}{3}\sin x.
\end{aligned}
$$

Since the domain of f is

$$-\frac{1}{3} \leq x \leq \frac{1}{3}$$

we obtain

$$
\begin{aligned}
-1 \leq 3x &\leq 1 \\
-\pi/2 \leq \sin^{-1} 3x &\leq \pi/2.
\end{aligned}
$$

Then the range of f is $[-\pi/2, \pi/2]$. Thus, the domain and range of f^{-1} are $[-\pi/2, \pi/2]$ and $[-1/3, 1/3]$, respectively.

81. To find f^{-1}, interchange x and y, solve for y, and replace y by $f^{-1}(x)$:

$$
\begin{aligned}
x &= \sin^{-1}(y/2) + 3 \\
\sin(x - 3) &= y/2 \\
2\sin(x - 3) &= y \\
f^{-1}(x) &= 2\sin(x - 3).
\end{aligned}
$$

Since the domain of f is

$$-2 \leq x \leq 2$$

we obtain

$$
\begin{aligned}
-1 \leq x/2 &\leq 1 \\
-\pi/2 \leq \sin^{-1}(x/2) &\leq \pi/2 \\
3 - \pi/2 \leq \sin^{-1}(x/2) + 3 &\leq 3 + \pi/2.
\end{aligned}
$$

Then the range of f is $[3 - \pi/2, 3 + \pi/2]$. Thus, the domain and range of f^{-1} are $[3 - \pi/2, 3 + \pi/2]$ and $[-2, 2]$, respectively.

83. Multiplying the equation by LCD, we get

$$
\begin{aligned}
13.7 \sin 33.2^\circ &= a \cdot \sin 45.6^\circ \\
\frac{13.7 \sin 33.2^\circ}{\sin 45.6^\circ} &= a \\
10.5 &\approx a.
\end{aligned}
$$

The solution set is $\{10.5^\circ\}$.

85. Since

$$\sin x = \frac{8.5 \sin(\pi/7)}{6.3}$$

we have

$$x = \sin^{-1}\left(\frac{8.5 \sin(\pi/7)}{6.3}\right) \approx 0.63 + 2k\pi$$

or

$$x = \pi - \sin^{-1}\left(\frac{8.5 \sin(\pi/7)}{6.3}\right) = 2.52 + 2k\pi$$

where k is an integer.

87. Since

$$\cos\alpha = \frac{5^2 + 6^2 - 7^2}{2(5)(6)}$$

we obtain

$$\alpha = \cos^{-1}\left(\frac{5^2 + 6^2 - 7^2}{2(5)(6)}\right) \approx 1.37 + 2k\pi$$

or

$$\alpha = 2\pi - \cos^{-1}\left(\frac{5^2 + 6^2 - 7^2}{2(5)(6)}\right) = 4.91 + 2k\pi$$

where k is an integer.

89. Since $c^2 = 19.34156...$, we get $c \approx \pm 4.40$.

91. Since

$$5\sin x = 2$$

we find

$$x = \sin^{-1}\left(\frac{2}{5}\right) \approx 0.41 + 2k\pi$$

or

$$x = \pi - \sin^{-1}\left(\frac{2}{5}\right) = 2.73 + 2k\pi$$

where k is an integer.

93. We find

$$\begin{aligned}
6\cos^{-1} x &= 3 \\
\cos^{-1} x &= \frac{1}{2} \\
x &= \cos\left(\frac{1}{2}\right) \\
x &\approx 0.88
\end{aligned}$$

95. Since $\sin(\pm\pi) = \sin 0 = 0$, we find that $\sin x = 0$ has three solutions in $(-2\pi, 2\pi)$. Since $\sin(\pi/2) = 1$, the maximum value of $y = \sin x$ in $(-2\pi, 2\pi)$ is 1.

97. a) $\pi/6$ **b)** $2\pi/3$ **c)** $-\pi/4$
d) $\sqrt{3}/2$ **e)** 0 **f)** $-\pi/2$

99. $(-\sin x)(\cos x)(-\tan x) =$

$$\sin x \cos x \frac{\sin x}{\cos x} = \sin^2 x$$

101. The amplitude is $A = 3$, period is
$\frac{2\pi}{B} = \frac{2\pi}{1} = 2\pi$, and phase shift is π.
The range is $[-3 + 9, 3 + 9]$ or $[6, 12]$.

103. The angle spanned by the first seventeen rectangles is

$$\tan^{-1}\left(\frac{1}{1}\right) + \tan^{-1}\left(\frac{1}{\sqrt{2}}\right) + ... + \tan^{-1}\left(\frac{1}{\sqrt{17}}\right) \approx 365°$$

while the angle spanned by the first sixteen rectangles is

$$\tan^{-1}\left(\frac{1}{1}\right) + \tan^{-1}\left(\frac{1}{\sqrt{2}}\right) + ... + \tan^{-1}\left(\frac{1}{\sqrt{16}}\right) \approx 351°.$$

Thus, the 17th rectangle is the first rectangle that overlaps with the first rectangle.

For Thought

1. False, rather $\alpha = \frac{3\pi}{2}$.

2. True

3. False, rather $x = \frac{\pi}{4} + \frac{k\pi}{2}$.

4. True, since $\cos(\pi/3) = \cos(5\pi/3) = 1/2$.

5. False, rather $5x = \frac{\pi}{4} + k\pi$.

6. True, since $\sin(\pi/4) = \sin(3\pi/4) = \sqrt{2}/2$.

7. True, since $\csc x = 1/\sin(x)$.

8. False, since $\cot^{-1} 3 = \tan^{-1}(1/3)$ and $\frac{1}{3} \neq 0.33$.

9. True, since $\cot^{-1}(3) = \tan^{-1}(1/3)$.

10. False, rather we have

$$\left\{x \mid 3x = \frac{\pi}{2} + 2k\pi\right\} = \left\{x \mid x = \frac{\pi}{6} + \frac{2k\pi}{3}\right\}.$$

4.3 Exercises

1. The values of $x/2$ in $[0, 2\pi)$ are $\pi/3$ and $5\pi/3$. Then we get

$$\frac{x}{2} = \frac{\pi}{3} + 2k\pi \text{ or } \frac{x}{2} = \frac{5\pi}{3} + 2k\pi$$

$$x = \frac{2\pi}{3} + 4k\pi \text{ or } x = \frac{10\pi}{3} + 4k\pi.$$

The solution set is

$$\left\{x \mid x = \frac{2\pi}{3} + 4k\pi \text{ or } x = \frac{10\pi}{3} + 4k\pi\right\}.$$

3. Value of $3x$ in $[0, 2\pi)$ is 0. Thus, $3x = 2k\pi$.

The solution set is $\left\{x \mid x = \frac{2k\pi}{3}\right\}$.

5. Since $\sin(x/2) = 1/2$, values of $x/2$ in $[0, 2\pi)$ are $\pi/6$ and $5\pi/6$. Then

$$\frac{x}{2} = \frac{\pi}{6} + 2k\pi \text{ or } \frac{x}{2} = \frac{5\pi}{6} + 2k\pi$$

$$x = \frac{\pi}{3} + 4k\pi \text{ or } x = \frac{5\pi}{3} + 4k\pi.$$

The solution set is

$$\left\{ x \mid x = \frac{\pi}{3} + 4k\pi \text{ or } x = \frac{5\pi}{3} + 4k\pi \right\}.$$

7. Since $\sin(2x) = -\sqrt{2}/2$, values of $2x$ in $[0, 2\pi)$ are $5\pi/4$ and $7\pi/4$. Thus,

$$2x = \frac{5\pi}{4} + 2k\pi \text{ or } 2x = \frac{7\pi}{4} + 2k\pi$$

$$x = \frac{5\pi}{8} + k\pi \text{ or } x = \frac{7\pi}{8} + k\pi.$$

The solution set is

$$\left\{ x \mid x = \frac{5\pi}{8} + k\pi \text{ or } x = \frac{7\pi}{8} + k\pi \right\}.$$

9. Value of $2x$ in $[0, \pi)$ is $\pi/3$. Then

$$2x = \frac{\pi}{3} + k\pi.$$

The solution set is $\left\{ x \mid x = \frac{\pi}{6} + \frac{k\pi}{2} \right\}.$

11. Value of $4x$ in $[0, \pi)$ is 0. Then

$$4x = k\pi.$$

The solution set is $\left\{ x \mid x = \frac{k\pi}{4} \right\}.$

13. The values of πx in $[0, 2\pi)$ are $\pi/6$ and $5\pi/6$. Then

$$\pi x = \frac{\pi}{6} + 2k\pi \text{ or } \pi x = \frac{5\pi}{6} + 2k\pi$$

$$x = \frac{1}{6} + 2k \text{ or } x = \frac{5}{6} + 2k.$$

The solution set is

$$\left\{ x \mid x = \frac{1}{6} + 2k \text{ or } x = \frac{5}{6} + 2k \right\}.$$

15. Values of $2\pi x$ in $[0, 2\pi)$ are $\pi/2$ and $3\pi/2$. So

$$2\pi x = \frac{\pi}{2} + 2k\pi \text{ or } 2\pi x = \frac{3\pi}{2} + 2k\pi$$

$$x = \frac{1}{4} + k \text{ or } x = \frac{3}{4} + k.$$

The solution set is

$$\left\{ x \mid x = \frac{1}{4} + \frac{k}{2} \right\}.$$

17. Set and solve

$$x + \frac{\pi}{4} = \frac{5\pi}{6}, \frac{13\pi}{6}.$$

The solution set is $\left\{ \frac{7\pi}{12}, \frac{23\pi}{12} \right\}.$

19. Set and solve

$$x - \frac{\pi}{8} = 0.$$

The solution set is $\left\{ \frac{\pi}{8} \right\}.$

21. First, $\sin(x - \pi) = -1/2$. Set and solve

$$x - \pi = -\frac{5\pi}{6}, -\frac{\pi}{6}.$$

The solution set is $\left\{ \frac{\pi}{6}, \frac{5\pi}{6} \right\}.$

23. First, $\tan(x - \pi/2) = 1$. Set and solve

$$x - \frac{\pi}{2} = \frac{\pi}{4}, \frac{5\pi}{4}.$$

The solution set is $\left\{ \frac{3\pi}{4}, \frac{7\pi}{4} \right\}.$

25. Since $\sin \alpha = -\sqrt{3}/2$, the solution set is $\{240°, 300°\}$.

27. Since $\sin 2x = -\sqrt{3}/2$, we obtain

$$2x = 240° + k \cdot 360°$$

or

$$2x = 300° + k \cdot 360°.$$

Then

$$x = 120° + k \cdot 180°$$

or

$$x = 150° + k \cdot 180°.$$

If $k = 0, 1$, then $x = 120°, 150°, 300°, 330°$.

29. Since $\cos 2x = -1/2$, we find

$$2x = 120° + k \cdot 360°$$

or
$$2x = 240° + k \cdot 360°.$$

Then
$$x = 60° + k \cdot 180°$$

or
$$x = 120° + k \cdot 180°.$$

If $k = 0, 1$, then $x = 60°, 120°, 240°, 300°$.

31. Since $\cos 2\alpha = 1/\sqrt{2}$, values of 2α in $[0, 360°)$ are $45°$ and $315°$. Thus,

$$2\alpha = 45° + k \cdot 360° \text{ or } 2\alpha = 315° + k \cdot 360°$$

$$\alpha = 22.5° + k \cdot 180° \text{ or } \alpha = 157.5° + k \cdot 180°.$$

Then let $k = 0, 1$. The solution set is

$$\{22.5°, 157.5°, 202.5°, 337.5°\}.$$

33. Values of 3α in $[0, 360°)$ are $135°$ and $225°$. Then

$$3\alpha = 135° + k \cdot 360° \text{ or } 3\alpha = 225° + k \cdot 360°$$

$$\alpha = 45° + k \cdot 120° \text{ or } \alpha = 75° + k \cdot 120°.$$

By choosing $k = 0, 1, 2$, one obtains the solution set $\{45°, 75°, 165°, 195°, 285°, 315°\}$.

35. The value of $\alpha/2$ in $[0, 180°)$ is $30°$. Then

$$\frac{\alpha}{2} = 30° + k \cdot 180°$$

$$\alpha = 60° + k \cdot 360°.$$

By choosing $k = 0$, the solution set is $\{60°\}$.

37. Since $\csc(4\alpha) = \sqrt{2}$, the values of 4α in $[0, 360°)$ are $45°$ and $135°$. Then

$$4\alpha = 45° + k \cdot 360° \text{ or } 4\alpha = 135° + k \cdot 360°$$

$$\alpha = 11.25° + k \cdot 90° \text{ or } \alpha = 33.75° + k \cdot 90°.$$

Choosing $k = 0, 1, 2, 3$, the solution set is
$\{11.25°, 33.75°, 101.25°, 123.75°, 191.25°, 213.75°,$
$281.25°, 303.75°\}$.

39. Since $\sin 2x = \sqrt{3}/2$, we obtain

$$2x = \frac{\pi}{3} + k \cdot 2\pi$$

or
$$2x = \frac{2\pi}{3} + k \cdot 2\pi$$

where k is an integer. Then

$$x = \frac{\pi}{6} + k \cdot \pi$$

or
$$x = \frac{\pi}{3} + k \cdot \pi$$

If $k = 0, 1$, then $x = \dfrac{\pi}{6}, \dfrac{\pi}{3}, \dfrac{7\pi}{6}, \dfrac{4\pi}{3}$.

41. Since $\cos 2x = 1/2$, we find

$$2x = \frac{\pi}{3} + k \cdot 2\pi$$

or
$$2x = \frac{5\pi}{3} + k \cdot 2\pi$$

where k is an integer. Then

$$x = \frac{\pi}{6} + k \cdot \pi$$

or
$$x = \frac{5\pi}{6} + k \cdot \pi$$

If $k = 0, 1$, then

$$x = \frac{\pi}{6}, \frac{5\pi}{6}, \frac{7\pi}{6}, \frac{11\pi}{6}.$$

43. Since $\tan 3x = 1$, we find

$$3x = \frac{\pi}{4} + k \cdot \pi.$$

where k is an integer. Then

$$x = \frac{\pi}{12} + \frac{k\pi}{3}.$$

If $k = 0, 1, ..., 5$, then

$$x = \frac{\pi}{12}, \frac{5\pi}{12}, \frac{9\pi}{12}, \frac{13\pi}{12}, \frac{17\pi}{12}, \frac{21\pi}{12}.$$

45. Since $\sin(x/3) = 1/\sqrt{2}$, we obtain

$$\frac{x}{3} = \frac{\pi}{4} + k \cdot 2\pi$$

or

$$\frac{x}{3} = \frac{3\pi}{4} + k \cdot 2\pi$$

where k is an integer. Then

$$x = \frac{3\pi}{4} + k \cdot 6\pi$$

or

$$x = \frac{9\pi}{4} + k \cdot 6\pi$$

Thus, $x = \frac{3\pi}{4}$.

47. Since $\tan(x/2) = 1/\sqrt{3}$, we obtain

$$\frac{x}{2} = \frac{\pi}{6} + k \cdot \pi$$

where k is an integer. Then

$$x = \frac{\pi}{3} + k \cdot 2\pi.$$

If $k = 0$, then $x = \frac{\pi}{3}$.

49. A solution is $3\alpha = \sin^{-1}(0.34) \approx 19.88°$. Another solution is $3\alpha = 180° - 19.88° = 160.12°$. Then

$$3\alpha = 19.88° + k \cdot 360° \text{ or } 3\alpha = 160.12° + k \cdot 360°$$

$$\alpha \approx 6.6° + k \cdot 120° \text{ or } \alpha \approx 53.4° + k \cdot 120°.$$

Solution set is

$$\{\alpha \mid \alpha = 6.6° + k \cdot 120° \text{ or } \alpha = 53.4° + k \cdot 120°\}.$$

51. A solution is $3\alpha = \sin^{-1}(-0.6) \approx -36.87°$. This is coterminal with $323.13°$. Another solution is $3\alpha = 180° + 36.87° = 216.87°$. Then

$$3\alpha = 323.13° + k \cdot 360° \text{ or } 3\alpha = 216.87° + k \cdot 360°$$

$$\alpha \approx 107.7° + k \cdot 120° \text{ or } \alpha \approx 72.3° + k \cdot 120°.$$

The solution set is

$$\{\alpha \mid \alpha = 72.3° + k120° \text{ or } \alpha = 107.7° + k120°\}.$$

53. A solution is $2\alpha = \cos^{-1}(1/4.5) \approx 77.16°$. Another solution is $2\alpha = 360° - 77.16° = 282.84°$. Thus,

$$2\alpha = 77.16° + k \cdot 360° \text{ or } 2\alpha = 282.84° + k \cdot 360°$$

$$\alpha \approx 38.6° + k \cdot 180° \text{ or } \alpha \approx 141.4° + k \cdot 180°.$$

The solution set is

$$\{\alpha \mid \alpha = 38.6° + k180° \text{ or } \alpha = 141.4° + k180°\}.$$

55. A solution is $\alpha/2 = \sin^{-1}(-1/2.3) \approx -25.77°$. This is coterminal with $334.23°$. Another solution is $\alpha/2 = 180° + 25.77° = 205.77°$. Thus,

$$\frac{\alpha}{2} = 334.23° + k \cdot 360° \text{ or } \frac{\alpha}{2} = 205.77° + k \cdot 360°$$

$$\alpha \approx 668.5° + k \cdot 720° \text{ or } \alpha \approx 411.5° + k \cdot 720°.$$

The solution set is

$$\{\alpha \mid \alpha = 668.5° + k720° \text{ or } \alpha = 411.5° + k720°\}.$$

57. Note, $5x = \sin^{-1}(1/3) + 2k\pi$ or $5x = \pi - \sin^{-1}(1/3) + 2k\pi$. Solving for x, we get $x = \dfrac{\sin^{-1}(1/3)}{5} + \dfrac{2k\pi}{5}$

or $x = \dfrac{\pi - \sin^{-1}(1/3)}{5} + \dfrac{2k\pi}{5}$.

Since $\dfrac{\sin^{-1}(1/3)}{5} \approx 0.07$ and

$\dfrac{\pi - \sin^{-1}(1/3)}{5} \approx 0.56$, the solution set is

$$\left\{x \mid x = 0.07 + \frac{2k\pi}{5}, x = 0.56 + \frac{2k\pi}{5}\right\}.$$

59. Note, $x/2 = \sin^{-1}(-6/10) + 2\pi + 2k\pi$ or $x/2 = \pi + \sin^{-1}(6/10) + 2k\pi$. Solving for x, we get $x = 2\left(\sin^{-1}(-6/10) + 2\pi\right) + 4k\pi$

or $x = 2\left(\pi + \sin^{-1}(6/10)\right) + 4k\pi$.

Since $2\left(\sin^{-1}(-6/10) + 2\pi\right) \approx 11.28$ and

$2\left(\pi + \sin^{-1}(6/10)\right) \approx 7.57$, the solution set is

$$\{x \mid x = 7.57 + 4k\pi \text{ or } x = 11.28 + 4k\pi\}.$$

61. Note, $\pi x = \cos^{-1}(2/9) + 2k\pi$ or
$\pi x = 2\pi - \cos^{-1}(2/9) + 2k\pi$. Solving

for x, we find $x = \dfrac{\cos^{-1}(2/9)}{\pi} + 2k$

or $x = \dfrac{2\pi - \cos^{-1}(2/9)}{\pi} + 2k$.

Since $\dfrac{\cos^{-1}(2/9)}{\pi} \approx 0.43$ and

$\dfrac{2\pi - \cos^{-1}(2/9)}{\pi} \approx 1.57$, the solution set is

$$\{x \mid x = 0.43 + 2k \text{ or } x = 1.57 + 2k\}.$$

63. Note, $\pi x - 1 = \tan^{-1}(3) + k\pi$. Solving for x,

we obtain $x = \dfrac{1 + \tan^{-1}(3)}{\pi} + k$.

Since $\dfrac{1 + \tan^{-1}(3)}{\pi} \approx 0.72$, the solution set is

$$\{x \mid x = 0.72 + k\}.$$

65. First, $\cos 2x = -1$. Then

$$2x = (2k + 1)\pi.$$

The solution set is $\left\{\dfrac{\pi}{2}, \dfrac{3\pi}{2}\right\}$.

67. Apply the double-angle identity for sine.

$$
\begin{aligned}
2\sin x \cos x - \cos x &= 0 \\
\cos x (2\sin x - 1) &= 0
\end{aligned}
$$

Then $\cos x = 0$ or $\sin x = 1/2$.

The solution set is $\left\{\dfrac{\pi}{6}, \dfrac{5\pi}{6}, \dfrac{\pi}{2}, \dfrac{3\pi}{2}\right\}$.

69. First, $\sin x = \dfrac{\sin x}{\cos x}$ or $\dfrac{\sin x(\cos x - 1)}{\cos x} = 0$

Then $\sin x = 0$ or $\cos x = 1$.

The solution set is $\{0, \pi, 2\pi\}$.

71. Since $\cot^2 x + 1 = \csc^2 x$, the given equation is equivalent to $\csc x = \cos x$. Then $\sin x \cos x = 1$ or $\sin 2x = 2$.

No solution, and the solution set is $\emptyset$.

73. First, $\cos x = \dfrac{\cos x}{\sin x}$ or $\dfrac{\cos x(\sin x - 1)}{\sin x} = 0$

Then $\cos x = 0$ or $\sin x = 1$.

The solution set is $\left\{\dfrac{\pi}{2}, \dfrac{3\pi}{2}\right\}$.

75. Apply a sum-to-product identity. The original equation is equivalent to

$$2\sin 2x \cos x = 0.$$

Then $\sin x = 0$ or $\cos x = 0$.

The solution set is $\left\{0, \dfrac{\pi}{2}, \pi, \dfrac{3\pi}{2}, 2\pi\right\}$.

77. Apply a sum identity for cosine. The original equation is equivalent to

$$\cos x = \dfrac{\sqrt{3}}{2}.$$

The solution set is $\left\{\dfrac{\pi}{6}, \dfrac{11\pi}{6}\right\}$.

79. Divide both sides by $1 - \tan x \tan 2x$.

$$\dfrac{\tan x + \tan 2x}{1 - \tan x \tan 2x} = 1.$$

Apply a sum identity for tangent. Then $\tan 3x = 1$ and $3x = \dfrac{\pi}{4} + k\pi$.

$$x = \dfrac{\pi(1 + 4k)}{12}.$$

Extraneous roots are $x = 3\pi/4, 7\pi/4$.

The solution set is $\left\{\dfrac{\pi}{12}, \dfrac{5\pi}{12}, \dfrac{13\pi}{12}, \dfrac{17\pi}{12}\right\}$.

81. Given below is the graph of
$y = \sin(x/2) - \cos(3x)$.
The intercepts or solutions on $[0, 2\pi)$ are
approximately $\{0.4, 1.9, 2.2, 4.0, 4.4, 5.8\}$.

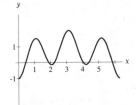

83. The graph of $y = \dfrac{x}{2} - \dfrac{\pi}{6} + \dfrac{\sqrt{3}}{2} - \sin x$

is shown. The solution set is $\{\pi/3\}$.

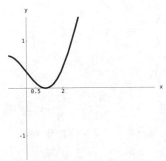

85. Since $v_o = 325$ and $d = 3300$, we have

$$
\begin{aligned}
325^2 \sin 2\theta &= 32(3300) \\
\sin 2\theta &= \frac{32(3300)}{325^2} \\
2\theta &= \sin^{-1}\left(\frac{32(3300)}{325^2}\right) \\
2\theta &\approx 88.74° \\
\theta &\approx 44.4°.
\end{aligned}
$$

Another angle is given by $2\theta = 180° - 88.74°$ $= 91.26°$ or $\theta = 91.26°/2 \approx 45.6°$. The muzzle was aimed at $44.4°$ or $45.6°$.

87. Note, 90 mph$= 90 \cdot \dfrac{5280}{3600}$ ft/sec$= 132$ ft/sec.

In $v_o^2 \sin 2\theta = 32d$, let $v_o = 132$ and $d = 230$.

$$
\begin{aligned}
132^2 \sin 2\theta &= 32(230) \\
2\theta &= \sin^{-1}\left(\frac{32(230)}{132^2}\right) \\
2\theta &\approx 25.0°
\end{aligned}
$$

Another possibility is $2\theta \approx 180° - 25° = 155°$. The two possible angles are

$$\theta \approx 12.5°, 77.5°.$$

The time it takes the ball to reach home plate can be found by using $x = v_o t \cos\theta$. (See Example 5). For the angle $12.5°$, it takes

$$t = \frac{230}{132 \cos 12.5°} \approx 1.78 \text{ sec}$$

while for $77.5°$ it takes

$$t = \frac{230}{132 \cos 77.5°} \approx 8.05 \text{ sec.}$$

The difference in time is $8.05 - 1.78 \approx 6.3$ sec.

89. First, find the values of t when $x = \sqrt{3}$.

$$
\begin{aligned}
2\sin\left(\frac{\pi t}{3}\right) &= \sqrt{3} \\
\sin\left(\frac{\pi t}{3}\right) &= \frac{\sqrt{3}}{2} \\
\frac{\pi t}{3} = \frac{\pi}{3} + 2k\pi \quad &\text{or} \quad \frac{\pi t}{3} = \frac{2\pi}{3} + 2k\pi \\
\pi t = \pi + 6k\pi \quad &\text{or} \quad \pi t = 2\pi + 6k\pi \\
t = 1 + 6k \quad &\text{or} \quad t = 2 + 6k
\end{aligned}
$$

Then the ball is $\sqrt{3}$ ft above sea level for the values of t satisfying

$$1 + 6k < t < 2 + 6k$$

where k is a nonnegative integer.

91. Infinitely many solutions

93. Suppose $y < 0$. Then

$$\text{arccot}\, y = \arctan\left(\frac{1}{y}\right) + \pi$$

Applying the addition rule for cosine, we find

$$
\begin{aligned}
\cos(\text{arccot}\, y) &= -\cos\left(\arctan\left(\frac{1}{y}\right)\right) \\
&= -\frac{1}{\sec\left(\arctan\left(\frac{1}{y}\right)\right)} \\
&= -\frac{1}{\sqrt{1 + \tan^2\left(\arctan\left(\frac{1}{y}\right)\right)}}
\end{aligned}
$$

since $\sec^2 x = 1 + \tan^2 x$. Since $\tan(\arctan x) = x$, we obtain

$$
\begin{aligned}
\cos(\text{arccot}\, y) &= -\frac{1}{\sqrt{1 + \frac{1}{y^2}}} \\
&= -\frac{\sqrt{y^2}}{\sqrt{y^2 + 1}} \\
\cos(\text{arccot}\, y) &= \frac{y}{\sqrt{y^2 + 1}}
\end{aligned}
$$

for $y < 0$. Similarly, if $y > 0$, we have

$$\cos(\text{arccot}\, y) = \frac{y}{\sqrt{y^2 + 1}}$$

95. a) $-\dfrac{\pi}{6}$ **b)** $\dfrac{3\pi}{4}$

c) $\dfrac{\pi}{4}$ **d)** $\dfrac{\pi}{4}$

97. Since cosine is an even function, we obtain

$$\cos\frac{\pi}{12} = \sqrt{\frac{1 + \cos(\pi/6)}{2}} = \sqrt{\frac{1 + \sqrt{3}/2}{2}} =$$

$$\sqrt{\frac{2 + \sqrt{3}}{4}} = \frac{\sqrt{2 + \sqrt{3}}}{2}$$

99. $(30 + 25)^2 = 3025$

For Thought

1. False, only solutions are $60°$ and $300°$.

2. False, since there is no solution in $[0, \pi)$.

3. True, since $-30°$ and $330°$ are coterminal angles.

4. True   **5.**   True, since the right-side is a factorization of the left-side.

6. False, $x = 0$ is a solution to the first equation and not to the second equation.

7. False, $\cos^{-1} 2$ is undefined.   **8.**   True

9. False, $x = 3\pi/4$ is not a solution to the first equation but is a solution to the second equation.

10. False, since $x = 3\pi/2$ does not satisfy both equations.

4.4 Exercises

1. Since $\cos x = \pm 1$, the solution set is $\{0, \pi\}$.

3. Since $\sin x = \pm\dfrac{1}{2}$, the solution set is $\{\pi/6, 5\pi/6, 7\pi/6, 11\pi/6\}$.

5. Since $\sin 2x = \pm\dfrac{\sqrt{2}}{2}$, we obtain

$$2x = \frac{\pi}{4} + k\frac{\pi}{2}.$$

Then

$$x = \frac{\pi(2k + 1)}{8}.$$

Thus, the solution set is
$\{\frac{\pi}{8}, \frac{3\pi}{8}, \frac{5\pi}{8}, \frac{7\pi}{8}, \frac{9\pi}{8}, \frac{11\pi}{8}, \frac{13\pi}{8}, \frac{15\pi}{8}\}$.

7. Since $\tan x = \pm\dfrac{\sqrt{3}}{3}$, the solution set is
$\{\frac{\pi}{6}, \frac{5\pi}{6}, \frac{7\pi}{6}, \frac{11\pi}{6}\}$.

9. Set the right side to zero and factor:

$$\sin x(\sin x - 1) = 0$$

Then $\sin x = 0$ or $\sin x = 1$.
The solution set is $\{0, \frac{\pi}{2}, \pi\}$.

11. Factor as

$$\cos x(2\cos x + 1) = 0$$

Then $\cos x = 0$ or $\cos x = -1/2$.
The solution set is $\{\frac{\pi}{2}, \frac{2\pi}{3}, \frac{3\pi}{2}, \frac{4\pi}{3}\}$.

13. Factor as

$$\tan x(3\tan^2 x - 1) = 0$$

Then $\tan x = 0$ or $\tan x = \pm\sqrt{3}/3$.
The solution set is $\{0, \frac{\pi}{6}, \frac{5\pi}{6}, \pi, \frac{7\pi}{6}, \frac{11\pi}{6}\}$.

15. Since $1 - \cos^2 2x + \cos 2x = 1$, we have $\cos 2x(\cos 2x - 1) = 0$. Then $\cos 2x = 0, 1$. Consequently, $2x = 2k\pi$ or $2x = \frac{\pi}{2} + k\pi$. That is, , $x = k\pi$ or $x = \frac{(2k+1)\pi}{4}$.
The solution set is $\{0, \pi, \frac{\pi}{4}, \frac{3\pi}{4}, \frac{5\pi}{4}, \frac{7\pi}{4}\}$.

17. Factoring we have

$$(2\sin x - 1)(\sin x - 1) = 0.$$

Then $\sin x = \frac{1}{2}, 1$. The solution set is
$\{\frac{\pi}{6}, \frac{5\pi}{6}, \frac{\pi}{2}\}$.

19. Factoring we have

$$(2\cos x + 1)(\cos x + 1) = 0.$$

Then $\cos x = -\frac{1}{2}, -1$. The solution set is
$\{\frac{2\pi}{3}, \frac{4\pi}{3}, \pi\}$.

21. Factoring we have

$$(\sin x + 1)(3\sin x - 1) = 0.$$

Then $\sin x = -1, \frac{1}{3}$. Notice, $\sin^{-1}(1/3) \approx 0.3$ and $\pi - \sin^{-1}(1/3) \approx 2.8$ The solution set is
$\{\frac{3\pi}{2}, 0.3, 2.8\}$.

23. Factoring we have

$$(\cos 2x + 1)(3\cos 2x - 1) = 0.$$

Then $\cos 2x = -1, \frac{1}{3}$. Consequently, either
$2x = \pi + 2k\pi$,

$2x = \cos^{-1}(1/3) + 2k\pi$,

$2x = 2\pi - \cos^{-1}(1/3) + 2k\pi$.

The solution set is $\{\frac{\pi}{2}, \frac{3\pi}{2}, 0.6, 2.5, 3.8, 5.7\}$.

25. Factoring we have
$$\left(2\sin\frac{x}{2}-1\right)\left(3\sin\frac{x}{2}-1\right)=0.$$
Then $\sin\frac{x}{2}=\frac{1}{2},\frac{1}{3}$.
Consequently, either $\frac{x}{2}=\frac{\pi}{6}+2k\pi$,
$\frac{x}{2}=\frac{5\pi}{6}+2k\pi$,
$\frac{x}{2}=\sin^{-1}\frac{1}{3}+2k\pi$, or
$\frac{x}{2}=\pi-\sin^{-1}\frac{1}{3}+2k\pi$.
Let $k=0$. The solution set is $\{\frac{\pi}{3},\frac{5\pi}{3},0.7,5.6\}$.

27. Since $|\sin\theta|\le 1$ and by the quadratic formula, we find
$$\sin\theta=\frac{3-\sqrt{13}}{2}.$$
Then $360°+\sin^{-1}\left(\frac{3-\sqrt{13}}{2}\right)\approx 342.4°$ and $180°-\sin^{-1}\left(\frac{3-\sqrt{13}}{2}\right)\approx 197.6°$
The solution set is $\{197.6°,342.4°\}$.

29. Applying the quadratic formula, we find
$$\cos\theta=\frac{1\pm\sqrt{33}}{8}.$$
Then $\theta_1=\cos^{-1}\left(\frac{1+\sqrt{33}}{8}\right)\approx 32.5°$ and
$\theta_2=\cos^{-1}\left(\frac{1-\sqrt{33}}{8}\right)\approx 126.4°$.
The two other solutions are
$360°-\theta_1\approx 327.5°$ and
$360°-\theta_2\approx 233.6°$
The solution set is
$\{32.5°,126.4°,233.6°,327.5°\}$.

31. By the method of completing the square, we get
$$\begin{aligned}
\tan^2\theta-2\tan\theta &= 1\\
\tan^2\theta-2\tan\theta+1 &= 2\\
(\tan\theta-1)^2 &= 2\\
\tan\theta-1 &= \pm\sqrt{2}\\
\theta=\tan^{-1}(1+\sqrt{2}) &\quad\text{or}\quad \theta=\tan^{-1}(1-\sqrt{2})\\
\theta\approx 67.5° &\quad\text{or}\quad \theta=-22.5°.
\end{aligned}$$
Other solutions are $180°+67.5°=247.5°$, $180°-22.5°=157.5°$, and $180°+157.5°=337.5°$. The solution set is $\{67.5°,157.5°,247.5°,337.5°\}$.

33. Square both sides.
$$\begin{aligned}
\tan^2\theta &= \sec^2\theta-2\sqrt{3}\sec\theta+3\\
-1 &= -2\sqrt{3}\sec\theta+3\\
2\sqrt{3}\sec\theta &= 4\\
\sec\theta &= \frac{2}{\sqrt{3}}.
\end{aligned}$$
Then $\theta=30°,330°$. But, $\theta=30°$ is an extraneous root. The solution set is $\{330°\}$.

35. Square both sides and apply the double angle identity for sine.
$$\begin{aligned}
1-\sin 2\theta &= 1-\frac{\sqrt{3}}{2}\\
\sin 2\theta &= \frac{\sqrt{3}}{2}
\end{aligned}$$
The solution set is $\{30°,240°\}$.

37. Applying a difference identity, we obtain
$$\sin(\theta-45°)=\frac{0.8}{\sqrt{2}}$$
Then $\theta=45°+\sin^{-1}\left(\frac{0.8}{\sqrt{2}}\right)\approx 79.4°$, or
$\theta=45°+180°-\sin^{-1}\left(\frac{0.8}{\sqrt{2}}\right)\approx 190.6°$
The solution set is $\{79.4°,190.6°\}$.

39. Since $-\sin x=\sin x$, we get $0=2\sin x$ or $\sin x=0$. The solution set is $\{0,\pi\}$.

41. Since $-\tan x=\tan x$, we get $0=2\tan x$ or $\tan x=0$. The solution set is $\{0,\pi\}$.

43. Set the right-hand side to zero and factor.
$$\begin{aligned}
3\sin^2 x-\sin x &= 0\\
\sin x(3\sin x-1) &= 0
\end{aligned}$$
Set each factor to zero.
$$\begin{aligned}
\sin x=0 &\quad\text{or}\quad \sin x=1/3\\
x=0,\pi &\quad\text{or}\quad x=\sin^{-1}(1/3)\approx 0.3
\end{aligned}$$
Another solution to $\sin x=1/3$ is $x=\pi-0.3\approx 2.8$.
The solution set is $\{0,0.3,2.8,\pi\}$.

45. Since
$$\cos x(2\cos x - 3) = 0$$
we obtain
$$\cos x = 0$$
or
$$\cos x = \frac{3}{2}.$$

The latter equation has no solution. The solutions of $\cos x = 0$ are
$$x = \frac{\pi}{2} + k \cdot \pi.$$

Thus, the solution set is $\left\{\dfrac{\pi}{2}, \dfrac{3\pi}{2}\right\}$.

47. Set the right-hand side to zero and factor.
$$\begin{aligned}
2\cos^2 x + 3\cos x + 1 &= 0 \\
(2\cos x + 1)(\cos x + 1) &= 0
\end{aligned}$$

Set the factors to zero.
$$\begin{aligned}
\cos x = -1/2 \quad &\text{or} \quad \cos x = -1 \\
x = 2\pi/3, 4\pi/3 \quad &\text{or} \quad x = \pi
\end{aligned}$$

The solution set is $\{\pi, 2\pi/3, 4\pi/3\}$.

49. Squaring both sides of the equation, we obtain
$$\begin{aligned}
\tan^2 x &= \sec^2 x - 2\sqrt{3}\sec x + 3 \\
\sec^2 x - 1 &= \sec^2 x - 2\sqrt{3}\sec x + 3 \\
-4 &= -2\sqrt{3}\sec x \\
\sec x &= 2/\sqrt{3} \\
x &= \pi/6, 11\pi/6.
\end{aligned}$$

Checking $x = \pi/6$, one gets $\tan(\pi/6) = 1/\sqrt{3}$ and $\sec(\pi/6) - \sqrt{3} = 2/\sqrt{3} - \sqrt{3} = -1/\sqrt{3}$. Then $x = \pi/6$ is an extraneous root and the solution set is $\{11\pi/6\}$.

51. Square both sides of the equation.
$$\begin{aligned}
\sin^2 x + 2\sqrt{3}\sin x + 3 &= 27\cos^2 x \\
\sin^2 x + 2\sqrt{3}\sin x + 3 &= 27(1 - \sin^2 x) \\
28\sin^2 x + 2\sqrt{3}\sin x - 24 &= 0 \\
14\sin^2 x + \sqrt{3}\sin x - 12 &= 0
\end{aligned}$$

By the quadratic formula, we get
$$\begin{aligned}
\sin x &= \frac{-\sqrt{3} \pm \sqrt{675}}{28} \\
\sin x &= \frac{-\sqrt{3} \pm 15\sqrt{3}}{28} \\
\sin x &= \frac{\sqrt{3}}{2}, \frac{-4\sqrt{3}}{7}.
\end{aligned}$$
Thus,
$$x = \frac{\pi}{3}, \frac{2\pi}{3} \quad \text{or} \quad x = \sin^{-1}\left(\frac{-4\sqrt{3}}{7}\right)$$
$$x = \frac{\pi}{3}, \frac{2\pi}{3} \quad \text{or} \quad x \approx -1.427.$$

Checking $x = 2\pi/3$, one finds $\sin(2\pi/3) + \sqrt{3} = \sqrt{3}/2 + \sqrt{3}$ and $3\sqrt{3}\cos(2\pi/3)$ is a negative number. Then $x = 2\pi/3$ is an extraneous root.

An angle coterminal with -1.427 is $2\pi - 1.427 \approx 4.9$. In a similar way, one checks that $\pi + 1.427 \approx 4.568$ is an extraneous root. Thus, the solution set is $\{\pi/3, 4.9\}$.

53. Substitute $\cos^2 x = 1 - \sin^2 x$.
$$\begin{aligned}
5\sin^2 x - 2\sin x &= 1 - \sin^2 x \\
6\sin^2 x - 2\sin x - 1 &= 0
\end{aligned}$$

Apply the quadratic formula.
$$\begin{aligned}
\sin x &= \frac{2 \pm \sqrt{28}}{12} \\
\sin x &= \frac{1 \pm \sqrt{7}}{6}
\end{aligned}$$
Then
$$x = \sin^{-1}\left(\frac{1 + \sqrt{7}}{6}\right) \quad \text{or} \quad x = \sin^{-1}\left(\frac{1 - \sqrt{7}}{6}\right)$$
$$x \approx 0.653 \quad \text{or} \quad x \approx -0.278.$$

Another solution is $\pi - 0.653 \approx 2.5$. An angle coterminal with -0.278 is $2\pi - 0.278 \approx 6.0$. Another solution is $\pi + 0.278 \approx 3.4$. The solution set is $\{0.7, 2.5, 3.4, 6.0\}$.

55. Express the equation in terms of $\sin x$ and $\cos x$.
$$\begin{aligned}
\frac{\sin x}{\cos x} \cdot 2\sin x \cos x &= 0 \\
2\sin^2 x &= 0 \\
\sin x &= 0
\end{aligned}$$

Solution set is $\{0, \pi\}$.

57. Substitute the double-angle identity for $\sin x$.

$$2 \sin x \cos x - \sin x \cos x = \cos x$$
$$\sin x \cos x - \cos x = 0$$
$$\cos x (\sin x - 1) = 0$$
$$\cos x = 0 \quad \text{or} \quad \sin x = 1$$
$$x = \pi/2, 3\pi/2 \quad \text{or} \quad x = \pi/2$$

Solution set is $\{\pi/2, 3\pi/2\}$.

59. Use the sum identity for sine.

$$\sin(x + \pi/4) = 1/2$$
$$x + \frac{\pi}{4} = \frac{\pi}{6} + 2k\pi \quad \text{or} \quad x + \frac{\pi}{4} = \frac{5\pi}{6} + 2k\pi$$
$$x = \frac{-\pi}{12} + 2k\pi \quad \text{or} \quad x = \frac{7\pi}{12} + 2k\pi$$

By choosing $k = 1$ in the first case and $k = 0$ in the second case, one finds the solution set is $\{7\pi/12, 23\pi/12\}$.

61. Apply the difference identity for sine.

$$\sin(2x - x) = -1/2$$
$$\sin x = -1/2$$

The solution set is $\{7\pi/6, 11\pi/6\}$.

63. Since $4 \cdot 4^{2\sin^2 x} = 4^{3\sin x}$, we set the exponents equal to each other. Then

$$2 \sin^2 x + 1 = 3 \sin x$$
$$2 \sin^2 x - 3 \sin x + 1 = 0$$
$$(2 \sin x - 1)(\sin x - 1) = 0$$
$$\sin x = 1, \frac{1}{2}.$$

Thus, the solution set is $\{\pi/6, \pi/2, 5\pi/6\}$.

65. Use a half-angle identity for cosine and express the equation in terms of $\cos \theta$.

$$\frac{1 + \cos \theta}{2} = \frac{1}{\cos \theta}$$
$$\cos \theta + \cos^2 \theta = 2$$
$$\cos^2 \theta + \cos \theta - 2 = 0$$
$$(\cos \theta + 2)(\cos \theta - 1) = 0$$
$$\cos \theta = -2 \quad \text{or} \quad \cos \theta = 1$$
$$\text{no solution} \quad \text{or} \quad \theta = 0°$$

Solution set is $\{0°\}$.

67. Divide both sides of the equation by $2 \cos \theta$.

$$\frac{\sin \theta}{\cos \theta} = \frac{1}{2}$$
$$\tan \theta = 0.5$$
$$\theta = \tan^{-1}(0.5) \approx 26.6°$$

Another solution is $180° + 26.6° = 206.6°$. Solution set is $\{26.6°, 206.6°\}$.

69. Since

$$2 \sin \theta \cos \theta = 3 \cos \theta$$

we have

$$\sin \theta (2 \cos \theta - 3) = 0.$$

Then

$$\sin \theta = 0 \quad \text{or} \quad \cos \theta = \frac{3}{2}.$$

The latter equation has no solution. The solutions of $\sin \theta = 0$ are

$$x = k \cdot 180°.$$

Thus, the solution set is $\{0°, 180°\}$.

71. Express the equation in terms of $\sin 3\theta$.

$$\sin 3\theta = \frac{1}{\sin 3\theta}$$
$$\sin^2 3\theta = 1$$
$$\sin 3\theta = \pm 1$$

Then

$$3\theta = 90° + k \cdot 360° \quad \text{or} \quad 3\theta = 270° + k \cdot 360°$$
$$\theta = 30° + k \cdot 120° \quad \text{or} \quad \theta = 90° + k \cdot 120°.$$

By choosing $k = 0, 1, 2$, one finds that the solution set is

$$\{30°, 90°, 150°, 210°, 270°, 330°\}.$$

73. By the method of completing the square, we get

$$\tan^2 \theta - 2 \tan \theta = 1$$
$$\tan^2 \theta - 2 \tan \theta + 1 = 2$$
$$(\tan \theta - 1)^2 = 2$$
$$\tan \theta - 1 = \pm \sqrt{2}$$
$$\theta = \tan^{-1}(1 + \sqrt{2}) \quad \text{or} \quad \theta = \tan^{-1}(1 - \sqrt{2})$$
$$\theta \approx 67.5° \quad \text{or} \quad \theta = -22.5°.$$

Other solutions are $180° + 67.5° = 247.5°$, $180° - 22.5° = 157.5°$, and $180° + 157.5° = 337.5°$. The solution set is

$$\{67.5°, 157.5°, 247.5°, 337.5°\}.$$

75. Factor as a perfect square.

$$(3\sin\theta + 2)^2 = 0$$
$$\sin\theta = -2/3$$
$$\theta = \sin^{-1}(-2/3) \approx -41.8°$$

An angle coterminal with $-41.8°$ is $360° - 41.8° = 318.2°$. Another solution is $180° + 41.8° = 221.8°$. The solution set is $\{221.8°, 318.2°\}$.

77. Factoring, we find

$$3\sin x - \sin x \cos x + 6 - 2\cos x = 0$$
$$\sin x(3 - \cos x) + 2(3 - \cos x) = 0$$
$$(\sin x + 2)(3 - \cos x) = 0$$
$$\sin x = -2 \quad \text{or} \quad \cos x = 3.$$

Since

$$-1 \le \sin x, \cos x \le 1$$

we see that there are no solutions. The solution set is the empty set $\emptyset$.

79. By using the sum identity for tangent, we get

$$\tan(3\theta - \theta) = \sqrt{3}$$
$$2\theta = 60° + k \cdot 180°$$
$$\theta = 30° + k \cdot 90°.$$

By choosing $k = 1, 3$, one obtains that the solution set is $\{120°, 300°\}$. Note, $30°$ and $210°$ are not solutions.

81. Factoring, we get

$$(4\cos^2\theta - 3)(2\cos^2\theta - 1) = 0.$$

Then

$$\cos^2\theta = 3/4 \quad \text{or} \quad \cos^2\theta = 1/2$$
$$\cos\theta = \pm\sqrt{3}/2 \quad \text{or} \quad \cos\theta = \pm 1/\sqrt{2}.$$

The solution set is

$$\{30°, 45°, 135°, 150°, 210°, 225°, 315°, 330°\}.$$

83. Factoring, we obtain

$$(\sec^2\theta - 1)(\sec^2\theta - 4) = 0$$
$$\sec^2\theta = 1 \quad \text{or} \quad \sec^2\theta = 4$$
$$\sec\theta = \pm 1 \quad \text{or} \quad \sec\theta = \pm 2.$$

Solution set is $\{0°, 60°, 120°, 180°, 240°, 300°\}$.

85. Since $\cot x = \dfrac{\cos x}{\sin x}$, we get

$$\frac{\cos x}{\sin x} = \sin x$$
$$\sin^2 x = \cos x$$
$$1 - \cos^2 = \cos x$$
$$\cos^2 + \cos x - 1 = 0.$$

By the quadratic formula, we get

$$\cos x = \frac{-1 \pm \sqrt{5}}{2}. \text{ Note, } \left|\frac{-1 - \sqrt{5}}{2}\right| > 1$$

and let $a = \cos^{-1}\left(\dfrac{-1 + \sqrt{5}}{2}\right)$. Since

$a \approx 0.905$, $2\pi - a \approx 5.379$, and $\cot a \approx 0.786$, the points of intersection are

$$(0.905 + 2k\pi, 0.786) \text{ and } (5.379 + 2k\pi, -0.786)$$

87. Since $a = \sqrt{3}$ and $b = 1$, we obtain $r = \sqrt{\sqrt{3}^2 + 1^2} = 2$. If the terminal side of α goes through $(\sqrt{3}, 1)$, then $\tan\alpha = 1/\sqrt{3}$. Then one can choose $\alpha = \pi/6$ and $x = 2\sin(2t + \pi/6)$. The times when $x = 0$ are given by

$$\sin\left(2t + \frac{\pi}{6}\right) = 0$$
$$2t + \frac{\pi}{6} = k \cdot \pi$$
$$2t = -\frac{\pi}{6} + k \cdot \pi$$
$$t = -\frac{\pi}{12} + \frac{k \cdot \pi}{2}$$
$$t = -\frac{\pi}{12} + \frac{\pi}{2} + \frac{k \cdot \pi}{2}$$
$$t = \frac{5\pi}{12} + \frac{k \cdot \pi}{2}$$

where k is a nonnegative integer.

89. Note, $\dfrac{1 - \cos x}{x} = 0$ implies $\cos x = 1$
and $x \neq 0$. Thus, $x = \pm 2\pi, \pm 4\pi, \pm 6\pi$.
In the interval $(-2\pi, 2\pi)$, using a calculator, we get that the maximum value of $\dfrac{1 - \cos x}{x}$
is about 0.7246.

91. Since $\sin x = \pm 1$, we have $x = \dfrac{\pi}{2}, \dfrac{3\pi}{2}$.

93. a) Domain $[-1, 1]$, range $[-\pi/2, \pi/2]$

    b) Domain $[-1, 1]$, range $[0, \pi]$

    c) Domain $(-\infty, \infty)$, range $(-\pi/2, \pi/2)$

95. $\dfrac{\sin y}{\cos y} \cdot \cos y + \dfrac{1}{\sin y} \cdot \sin^2 y = 2 \sin y$

97. The quadrilateral has vertices $A(0,0)$, $B(0, -9/2)$, $C(82/17, -15/17)$, and $D(7/2, 0)$. The area of triangle $\triangle ABC$ is

$$\frac{1}{2} \cdot \frac{9}{2} \cdot \frac{82}{17} = \frac{369}{34}$$

and the area of triangle $\triangle ACD$ is

$$\frac{1}{2} \cdot \frac{7}{2} \cdot \frac{15}{17} = \frac{105}{68}.$$

The sum of areas of the two triangles is the area of the quadrilateral, i.e.,

$$\text{Area} = \frac{369}{34} + \frac{105}{68} = \frac{843}{68} \quad \text{square units.}$$

Review Exercises

1. $-\pi/6$

3. $-\pi/4$

5. $\pi/4$

7. $\cos(\arcsin(1/2) = \cos(\pi/6) = \sqrt{3}/2$

9. $\tan(\arcsin(\sqrt{2}/2) = \tan(\pi/4) = 1$

11. $\sin^{-1}(\sin(-\pi/4)) = \sin^{-1}(-\sqrt{2}/2) = -\pi/4$

13. $\sin^{-1}(\sin(3\pi/4)) = \sin^{-1}(\sqrt{2}/2) = \pi/4$

15. $\cos^{-1}(\cos(-\pi/6)) = \cos^{-1}(\sqrt{3}/2) = \pi/6$

17. $\csc^{-1}(\sec(\pi/3)) = \csc^{-1}(2) = \pi/6$

19. $90°$

21. $135°$

23. $30°$

25. $90°$

27. Since $\cos x = -1$, the solution set is $\{\pi\}$.

29. Since $\sin x = 1/2$, the solution set is $\{\pi/6, 5\pi/6\}$.

31. Since $\tan x = -1$, the solution set is $\{3\pi/4, 7\pi/4\}$.

33. Since $-2\sin(x) + 1 = 0$, we get $\sin x = 1/2$. The solution set is $\{30°, 150°\}$.

35. Since $2\cos(x) = \sqrt{2}$, we get $\cos x = \sqrt{2}/2$. The solution set is $\{45°, 315°\}$.

37. Since $-\sqrt{3}\tan(x) - 1 = 0$, we get $\tan x = -1/\sqrt{3}$. The solution set is $\{150°, 330°\}$.

39. Note, $a = \cos^{-1}(-3/5) \approx 2.21$ and $2\pi - a \approx 4.07$. The solution set is $\{2.21, 4.07\}$.

41. Since $2x = \sin^{-1}(2/9) + 2k\pi$ or $2x = \pi - \sin^{-1}(2/9) + 2k\pi$, we obtain

$$x = \frac{\sin^{-1}(2/9)}{2} + k\pi \quad \text{or}$$

$$x = \frac{\pi - \sin^{-1}(2/9)}{2} + k\pi.$$

By letting $k = 0, 1$, we find that the solution set is $\{0.11, 1.46, 3.25, 4.60\}$.

43. Note,

$$a = \cot^{-1}(\sqrt{2}) = \tan^{-1}(1/\sqrt{2}) \approx 0.62$$

and

$$\pi + a \approx 3.78.$$

The solution set is $\{0.62, 3.76\}$.

45. Solving for t, we find

$$\sin t = \frac{x}{3}$$

$$t = \sin^{-1}\left(\frac{x}{3}\right)$$

47. Solving for y, we obtain

$$\sin(2y) = \frac{a}{3}$$
$$2y = \sin^{-1}\left(\frac{a}{3}\right)$$
$$y = \frac{1}{2}\sin^{-1}\left(\frac{a}{3}\right)$$

49. Solving for h, we get

$$-2\cos h = q - 1$$
$$\cos h = \frac{q-1}{-2}$$
$$\cos h = \frac{1-q}{2}$$
$$h = \cos^{-1}\left(\frac{1-q}{2}\right)$$

51. Solving for x, we find

$$5\tan(\pi x) = b + 3$$
$$\tan(\pi x) = \frac{b+3}{5}$$
$$\pi x = \tan^{-1}\left(\frac{b+3}{5}\right)$$
$$x = \frac{1}{\pi}\tan^{-1}\left(\frac{b+3}{5}\right)$$

53. Solving for y, we obtain

$$\cos^{-1}(y-2) = \frac{x}{a}$$
$$y - 2 = \cos\left(\frac{x}{a}\right)$$
$$y = \cos\left(\frac{x}{a}\right) + 2$$

55. To find f^{-1}, interchange x and y, solve for y, and replace y by $f^{-1}(x)$:

$$x = \sin y$$
$$\sin^{-1} x = y$$
$$f^{-1}(x) = \sin^{-1} x.$$

Since the domain of f is

$$-\pi/2 \le x \le \pi/2$$

we obtain

$$-1 \le \sin x \le 1.$$

Then the range of f is $[-1, 1]$. Thus, the domain and range of f^{-1} are $[-1, 1]$ and $[-\pi/2, \pi/2]$, respectively.

57. To find f^{-1}, interchange x and y, solve for y, and replace y by $f^{-1}(x)$:

$$x = \sin^{-1} y$$
$$\sin x = y$$
$$f^{-1}(x) = \sin x.$$

Since the domain of f is

$$-1 \le x \le 1$$

we find

$$-\pi/2 \le \sin^{-1} x \le \pi/2.$$

Then the range of f is $[-\pi/2, \pi/2]$. Thus, the domain and range of f^{-1} are $[-\pi/2, \pi/2]$ and $[-1, 1]$, respectively.

59. Interchange x and y, solve for y, and replace y by $f^{-1}(x)$:

$$x = \sin 3y$$
$$\sin^{-1} x = 3y$$
$$\frac{1}{3}\sin^{-1} x = y$$
$$f^{-1}(x) = \frac{1}{3}\sin^{-1} x.$$

Since the domain of f is

$$-\pi/6 \le x \le \pi/6$$

we obtain

$$-\pi/2 \le 3x \le \pi/2$$
$$-1 \le \sin 3x \le 1.$$

Then the range of f is $[-1, 1]$. Thus, the domain and range of f^{-1} are $[-1, 1]$ and $[-\pi/6, \pi/6]$, respectively.

61. Interchange x and y, solve for y, and replace y by $f^{-1}(x)$:

$$
\begin{aligned}
x &= 6\cos 4y \\
\frac{x}{6} &= \cos 4y \\
\cos^{-1}\left(\frac{x}{6}\right) &= 4y \\
\frac{1}{4}\cos^{-1}\left(\frac{x}{6}\right) &= y \\
f^{-1}(x) &= \frac{1}{4}\cos^{-1}\left(\frac{x}{6}\right).
\end{aligned}
$$

Since the domain of f is

$$0 \le x \le \pi/4$$

we obtain

$$
\begin{aligned}
0 \le 4x &\le \pi \\
-1 \le \cos 4x &\le 1 \\
-6 \le 6\cos 4x &\le 6.
\end{aligned}
$$

Then the range of f is $[-6, 6]$. Thus, the domain and range of f^{-1} are $[-6, 6]$ and $[0, \pi/4]$, respectively.

63. Interchange x and y, solve for y, and replace y by $f^{-1}(x)$:

$$
\begin{aligned}
x &= 4 + \tan\left(\frac{\pi y}{2}\right) \\
x - 4 &= \tan\left(\frac{\pi y}{2}\right) \\
\tan^{-1}(x-4) &= \frac{\pi y}{2} \\
\frac{2}{\pi}\tan^{-1}(x-4) &= y \\
f^{-1}(x) &= \frac{2}{\pi}\tan^{-1}(x-4).
\end{aligned}
$$

Since the domain of f is

$$-1 < x < 1$$

we obtain

$$-\frac{\pi}{2} < \frac{\pi x}{2} < \frac{\pi}{2}$$

$$-\infty < \tan\left(\frac{\pi x}{2}\right) < \infty$$

$$-\infty < 4 + \tan\left(\frac{\pi x}{2}\right) < \infty.$$

Then the range of f is $(-\infty, \infty)$. Thus, the domain and range of f^{-1} are $(-\infty, \infty)$ and $(-1, 1)$, respectively.

65. $-\pi/6$

67. None, since cosine is negative in quadrants 2 and 3.

69. $-\pi/4$

71. $\pm\pi/4$

73. $-\pi/6$

75. Isolate $\cos 2x$ on one side.

$$
\begin{aligned}
2\cos 2x &= -1 \\
\cos 2x &= -\frac{1}{2}
\end{aligned}
$$

$$2x = \frac{2\pi}{3} + 2k\pi \quad\text{or}\quad 2x = \frac{4\pi}{3} + 2k\pi$$

$$x = \frac{\pi}{3} + k\pi \quad\text{or}\quad x = \frac{2\pi}{3} + k\pi$$

The solution set is

$$\left\{ x \mid x = \frac{\pi}{3} + k\pi \text{ or } x = \frac{2\pi}{3} + k\pi \right\}.$$

77. Set each factor to zero.

$$(\sqrt{3}\csc x - 2)(\csc x - 2) = 0$$

$$\csc x = \frac{2}{\sqrt{3}} \quad\text{or}\quad \csc x = 2$$

Thus, $x = \dfrac{\pi}{3}, \dfrac{2\pi}{3}, \dfrac{\pi}{6}, \dfrac{5\pi}{6}$ plus multiples of 2π. The solution set is

$$\left\{ x \mid x = \frac{\pi}{3} + 2k\pi, \frac{2\pi}{3} + 2k\pi, \frac{\pi}{6} + 2k\pi, \frac{5\pi}{6} + 2k\pi \right\}.$$

79. Set the right-hand side to zero and factor.

$$
\begin{aligned}
2\sin^2 x - 3\sin x + 1 &= 0 \\
(2\sin x - 1)(\sin x - 1) &= 0 \\
\sin x = \frac{1}{2} \quad\text{or}\quad \sin x &= 1
\end{aligned}
$$

The $x = \dfrac{\pi}{6}, \dfrac{5\pi}{6}, \dfrac{\pi}{2}$ plus multiples of 2π. The solution set is

$$\left\{ x \mid x = \frac{\pi}{6} + 2k\pi, \frac{5\pi}{6} + 2k\pi, \frac{\pi}{2} + 2k\pi \right\}.$$

81. Isolate $\sin \frac{x}{2}$ on one side.

$$\sin \frac{x}{2} = \frac{12}{8\sqrt{3}}$$
$$\sin \frac{x}{2} = \frac{3}{2\sqrt{3}}$$
$$\sin \frac{x}{2} = \frac{\sqrt{3}}{2}$$
$$\frac{x}{2} = \frac{\pi}{3} + 2k\pi \quad \text{or} \quad \frac{x}{2} = \frac{2\pi}{3} + 2k\pi$$
$$x = \frac{2\pi}{3} + 4k\pi \quad \text{or} \quad x = \frac{4\pi}{3} + 4k\pi$$

The solution set is

$$\left\{ x \mid x = \frac{2\pi}{3} + 4k\pi \text{ or } x = \frac{4\pi}{3} + 4k\pi \right\}.$$

83. By using the double-angle identity for sine, we get

$$\cos \frac{x}{2} - \sin \left(2 \cdot \frac{x}{2} \right) = 0$$
$$\cos \frac{x}{2} - 2 \sin \frac{x}{2} \cos \frac{x}{2} = 0$$
$$\cos \frac{x}{2} \left(1 - 2 \sin \frac{x}{2} \right) = 0$$
$$\cos \frac{x}{2} = 0 \quad \text{or} \quad \sin \frac{x}{2} = \frac{1}{2}.$$

Then $\frac{x}{2} = \frac{\pi}{2} + k\pi$, or $x = \frac{\pi}{6}, \frac{5\pi}{6}$ plus mutiples of 2π. Thus, $x = \pi + 2k\pi$ or $x = \frac{\pi}{3}, \frac{5\pi}{3}$ plus multiples of 4π. The solution set is

$$\left\{ x \mid x = \pi + 2k\pi, \frac{\pi}{3} + 4k\pi, \frac{5\pi}{3} + 4k\pi \right\}.$$

85. By the double-angle identity for cosine, we find

$$\cos 2x + \sin^2 x = 0$$
$$\cos^2 x - \sin^2 x + \sin^2 x = 0$$
$$\cos^2 x = 0$$
$$x = \frac{\pi}{2} + k\pi.$$

The solution set is $\left\{ x \mid x = \frac{\pi}{2} + k\pi \right\}.$

87. By factoring, we obtain

$$\sin x (\cos x + 1) + (\cos x + 1) = 0$$
$$(\sin x + 1)(\cos x + 1) = 0.$$

Then

$$\sin x = -1 \quad \text{or} \quad \cos x = -1$$
$$x = \frac{3\pi}{2} + 2k\pi \quad \text{or} \quad x = \pi + 2k\pi.$$

The solution set is

$$\left\{ x \mid x = \frac{3\pi}{2} + 2k\pi \text{ or } x = \pi + 2k\pi \right\}.$$

89. By multiplying the equation by 2, we obtain

$$2 \sin \alpha \cos \alpha = 1$$
$$\sin 2\alpha = 1$$
$$2\alpha = 90° + k360°$$
$$\alpha = 45° + k180°.$$

If $k = 0, 1$, then the solution set is

$$\{45°, 225°\}.$$

91. Suppose $1 + \cos \alpha \neq 0$. Dividing the equation by $1 + \cos \alpha$, we get

$$\frac{\sin \alpha}{1 + \cos \alpha} = 1$$
$$\tan \frac{\alpha}{2} = 1$$
$$\frac{\alpha}{2} = 45° + k180°$$
$$\alpha = 90° + k360°.$$

One solution is $90°$. On the other hand if $1 + \cos \alpha = 0$, then $\cos \alpha = -1$ and $\alpha = 180°$. Note $\alpha = 180°$ satisfies the given equation. The solution set is $\{90°, 180°\}.$

93. No solution since the left-hand side is equal to 1 by an identity. The solution set is $\emptyset$.

95. Isolate $\sin 2\alpha$ on one side.

$$\sin^4 2\alpha = \frac{1}{4}$$
$$\sin 2\alpha = \pm \sqrt[4]{\frac{1}{4}}$$
$$\sin 2\alpha = \pm \frac{1}{\sqrt{2}}$$
$$2\alpha = 45° + k90°$$
$$\alpha = 22.5° + k45°$$

By choosing $k = 0, 1, ..., 7$, one gets the solution set $\{22.5°, 67.5°, 112.5°, 157.5°, 202.5°, 247.5°, 292.5°, 337.5°\}.$

97. Suppose $\tan\alpha \neq 0$. Divide the equation by $\tan\alpha$.

$$\frac{2\tan\alpha}{1-\tan^2\alpha} = \tan\alpha$$

$$\frac{2}{1-\tan^2\alpha} = 1$$

$$2 = 1-\tan^2\alpha$$

$$\tan^2\alpha = -1$$

The last equation is inconsistent since $\tan^2\alpha$ is nonnegative. But if $\tan\alpha = 0$, then $\alpha = 0°, 180°$ and these two values of α satisfy the given equation.
The solution set is $\{0°, 180°\}$.

99. Using the sum identity for sine, we obtain

$$\sin(2\alpha + \alpha) = \cos 3\alpha$$

$$\sin 3\alpha = \cos 3\alpha$$

$$\tan 3\alpha = 1$$

$$3\alpha = 45° + k180°$$

$$\alpha = 15° + k60°.$$

By choosing $k = 0, 1, ..., 5$, we find the solution set

$$\{15°, 75°, 135°, 195°, 255°, 315°\}.$$

101. If $\sin x = \cos x$, then $\tan x = 1$ and $x = \dfrac{\pi}{4} + k\pi$ where k is an integer.
The points of intersection are

$$\left(\frac{\pi}{4} + 2k\pi, \frac{\sqrt{2}}{2}\right)$$

and

$$\left(\frac{5\pi}{4} + 2k\pi, -\frac{\sqrt{2}}{2}\right).$$

103. Let $a = 0.6$, $b = 0.4$, and

$$r = \sqrt{0.6^2 + 0.4^2} \approx 0.72.$$

If the terminal side of α goes through $(0.6, 0.4)$, then

$$\tan\alpha = 0.4/0.6$$

and we can choose

$$\alpha = \tan^{-1}(2/3) \approx 0.588.$$

Thus,

$$x = 0.72\sin(2t + 0.588).$$

The values of t when $x = 0$ are given by

$$\sin(2t + 0.588) = 0$$

$$2t + 0.588 = k\pi$$

$$2t = -0.588 + k\pi$$

$$t = -0.294 + \frac{k\pi}{2}$$

When $k = 1, 2$, we get

$$t \approx 1.28 \text{ sec}, 2.85 \text{ sec}.$$

105. Note, the number of handshakes in a group with n people is

$$1 + 2 + 3 + ... + (n-1) = \frac{(n-1)n}{2}.$$

In the first delegation, the number of handshakes is

$$\frac{(n-1)n}{2} = 190$$

which implies that $n = 20$, the number of members in the first delegation.

Since there were 480 handshakes between the first delegation and second delegation, the size of the second delegation is

$$\frac{480}{n} = \frac{480}{20} = 24 \text{ delegates}.$$

Thus, the total number of delegates is $20 + 24$, or 44 delegates.

Chapter 4 Test

1. $-\pi/6$ **2.** $2\pi/3$

3. $-\pi/4$ **4.** $\pi/6$

5. $-\pi/4$ **6.** $\sqrt{1 - \left(-\dfrac{1}{3}\right)^2} = \dfrac{2\sqrt{2}}{3}$

7. Since $-\sin\theta = 1$, we get $\sin\theta = -1$ and the solution set is $\left\{\theta \mid \theta = \dfrac{3\pi}{2} + 2k\pi\right\}$.

8. Since $\cos 3s = \dfrac{1}{2}$, we obtain

$$3s = \frac{\pi}{3} + 2k\pi \quad \text{or} \quad 3s = \frac{5\pi}{3} + 2k\pi$$
$$s = \frac{\pi}{9} + \frac{2k\pi}{3} \quad \text{or} \quad s = \frac{5\pi}{9} + \frac{2k\pi}{3}.$$

The solution set is

$$\left\{ s \mid s = \frac{\pi}{9} + \frac{2k\pi}{3} \text{ or } s = \frac{5\pi}{9} + \frac{2k\pi}{3} \right\}.$$

9. Since $\tan 2t = -\sqrt{3}$, we have

$$2t = \frac{2\pi}{3} + k\pi$$
$$t = \frac{\pi}{3} + \frac{k\pi}{2}.$$

The solution set is $\left\{ t \mid t = \dfrac{\pi}{3} + \dfrac{k\pi}{2} \right\}$.

10.

$$2\sin\theta \cos\theta = \cos\theta$$
$$\cos\theta(2\sin\theta - 1) = 0$$
$$\cos\theta = 0 \quad \text{or} \quad \sin\theta = 1/2$$

The solution set is

$$\left\{ \theta \mid \theta = \frac{\pi}{2} + k\pi, \frac{\pi}{6} + 2k\pi, \frac{5\pi}{6} + 2k\pi \right\}.$$

11. Since $\csc\alpha = \dfrac{8}{4}$, we get $\sin\alpha = \dfrac{1}{2}$.

The solution set is $\{30°, 150°\}$.

12. Since $\cot(\alpha/2) = -1$, we get $\dfrac{\alpha}{2} = 135°+k\cdot180°$ or $\alpha = 270°+k\cdot360°$ The solution set is $\{270°\}$.

13. Since $\sec\alpha = \dfrac{2}{\sqrt{3}}$, we obtain $\cos\alpha = \dfrac{\sqrt{3}}{2}$.

The solution set is $\{30°, 330°\}$.

14. Since $\sin\alpha = 0$ or $\cos\alpha = 0$, the solution set is $\{0°, 90°, 180°, 270°\}$.

15. Since $\sin(2\alpha) = \pm 1$, we find $2\alpha = 90°+k\cdot180°$ or $\alpha = 45° + k \cdot 90°$. Thus, the solution set is $\{45°, 135°, 225°, 315°\}$.

16. By factoring, we obtain

$$(3\sin\alpha - 1)(\sin\alpha - 1) = 0$$
$$\sin\alpha = 1/3 \quad \text{or} \quad \sin\alpha = 1$$
$$\alpha = \sin^{-1}(1/3) \approx 19.5° \quad \text{or} \quad \alpha = 90°.$$

Another solution is $\alpha = 180° - 19.5° = 160.5°$. The solution set is $\{19.5°, 90°, 160.5°\}$.

17.

$$\tan(2\alpha - 7\alpha) = 1$$
$$\tan(-5\alpha) = 1$$
$$-\tan 5\alpha = 1$$
$$\tan 5\alpha = -1$$
$$5\alpha = 135° + k180°$$
$$\alpha = 27° + k36°$$

The solution set is

$$\{27°, 63°, 99°, 171°, 207°, 243°, 279°, 351°\}.$$

Note, $135°$ and $315°$ are not solutions.

18. Solving for t, we get

$$\sin 2t = \frac{a}{5}$$
$$2t = \sin^{-1}\left(\frac{a}{5}\right)$$
$$t = \frac{1}{2}\sin^{-1}\left(\frac{a}{5}\right)$$

19. Interchange x and y, solve for y, and replace y by $f^{-1}(x)$:

$$x = \frac{1}{5}\cos\left(\frac{y}{4}\right) + \frac{1}{5}$$
$$5x = \cos\left(\frac{y}{4}\right) + 1$$
$$5x - 1 = \cos\left(\frac{y}{4}\right)$$
$$\cos^{-1}(5x - 1) = \frac{y}{4}$$
$$4\cos^{-1}(5x - 1) = y$$
$$f^{-1}(x) = 4\cos^{-1}(5x - 1).$$

Since the domain of f is

$$0 \le x \le 4\pi$$

we obtain

$$0 \le \frac{x}{4} \le \pi$$

$$-1 \le \cos\left(\frac{x}{4}\right) \le 1$$

$$-\frac{1}{5} \le \frac{1}{5}\cos\left(\frac{x}{4}\right) \le \frac{1}{5}$$

$$0 \le \frac{1}{5}\cos\left(\frac{x}{4}\right) + \frac{1}{5} \le \frac{2}{5}.$$

Then the range of f is $[0, 2/5]$. Thus, the domain and range of f^{-1} are $[0, 2/5]$ and $[0, 4\pi]$, respectively.

20. If $\cos x = \dfrac{1}{\cos x}$, then $\cos^2 x = 1$.

Thus, $x = k\pi$ where k is an integer. The points of intersection are $(2k\pi, 1)$ and $(\pi + 2k\pi, -1)$.

21. Let $a = 2$, $b = -4$, and

$$r = \sqrt{2^2 + (-4)^2} = \sqrt{20}.$$

If the terminal side of α goes through $(2, -4)$, then one can choose

$$\alpha = \tan^{-1}(-4/2) \approx -1.107.$$

Then

$$d = \sqrt{20}\sin(3t - 1.107).$$

The values of t when $d = 0$ are given by

$$\begin{aligned}
\sin(3t - 1.107) &= 0 \\
3t - 1.107 &= k\pi \\
3t &= 1.107 + k\pi \\
t &\approx 0.4 + \frac{k\pi}{3}.
\end{aligned}$$

By choosing $k = 0, 1, 2, 3$, we obtain the values of t in $[0, 4]$, namely, 0.4 sec, 1.4 sec, 2.5 sec, and 3.5 sec.

Tying It All Together

1. $\sqrt{2}/2$ **2.** $-1/2$

3. 1 **4.** $1/2$

5. $-\pi/6$ **6.** $2\pi/3$

7. $-\pi/4$ **8.** 0

9. $\sin(5\pi/3) = -\sqrt{3}/2$ **10.** $\cos(11\pi/6) = \sqrt{3}/2$

11. 1 **12.** 1

13. Period 2π, amplitude 1, phase shift $\pi/6$, and since the solutions of $\sin(x - \pi/6) = 0$ are $x = \pi/6 + k\pi$ it follows that the x-intercepts are $(\pi/6 + k\pi, 0)$ where k is an integer.

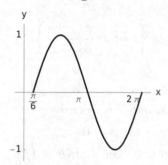

14. Period π, amplitude 1, phase shift 0, and since the solutions of $\sin(2x) = 0$ are $x = \dfrac{k\pi}{2}$ it follows that the x-intercepts are

$$\left(\frac{k\pi}{2}, 0\right)$$

where k is an integer.

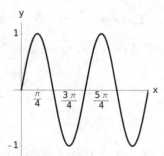

15. Period $2\pi/3$, amplitude 1, phase shift 0, and since the solutions of $\cos(3x) = 0$ are

$$x = \frac{\pi}{6} + \frac{k\pi}{3}$$ it follows that the x-intercepts are

$$\left(\frac{\pi}{6} + \frac{k\pi}{3}, 0\right)$$

where k is an integer.

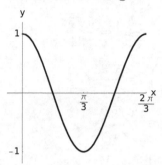

16. Note,

$$y = \cos\left(2\left(x - \frac{\pi}{2}\right)\right).$$

The period is π, amplitude is 1, and the phase shift is $\pi/2$. Since the solutions of

$$\cos(2x - \pi) = 0$$

are $x = \dfrac{\pi}{4} + \dfrac{k\pi}{2}$, the x-intercepts are

$$\left(\frac{\pi}{4} + \frac{k\pi}{2}, 0\right) \quad \text{or} \quad \left(\frac{3\pi}{4} + \frac{k\pi}{2}, 0\right)$$

where k is an integer.

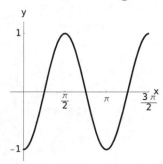

17. Note,

$$y = 3\sin\left(\frac{1}{2}\left(x - \pi\right)\right).$$

The period is 4π, amplitude is 3, and the phase shift is π. Since the solutions of

$$\sin\left(\frac{x}{2} - \frac{\pi}{2}\right) = 0$$

are $x = \pi + 2k\pi$, the x-intercepts are

$$(\pi + 2k\pi, 0)$$

where k is an integer.

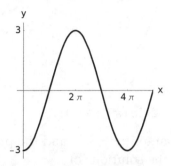

18. Period 2π, amplitude 4, phase shift $-\pi/3$, and since the solutions of

$$\cos(x + \pi/3) = 0$$

are $x = \pi/6 + k\pi$, the x-intercepts are

$$(\pi/6 + k\pi, 0)$$

where k is an integer.

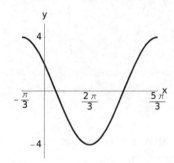

19. Note,

$$y = -4\cos\left(\frac{\pi}{3}x\right).$$

Then we obtain the following: period $\dfrac{2\pi}{\pi/3}$ or 6, amplitude 4, phase shift 0. Since the solutions of

$$\cos(\pi x/3) = 0$$

are $x = \dfrac{3}{2} + 3k$, the x-intercepts are

$$\left(\frac{3}{2} + 3k, 0\right)$$

where k is an integer.

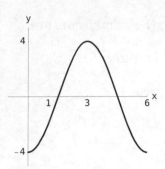

20. Note,
$$y = -2\sin\left(\pi\left(x - 1\right)\right).$$

The period is 2, amplitude is 2, and the phase shift is 1. Since the solutions of
$$\sin(\pi x - \pi) = 0$$
are $x = k$, the x-intercepts are
$$(k, 0)$$
where k is an integer.

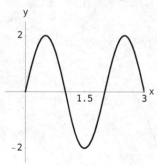

21. Since α is in Quadrant III, we get
$$\cos\alpha = -\sqrt{1 - \left(-\frac{1}{3}\right)^2} = -\frac{2\sqrt{2}}{3},$$
$$\csc\alpha = -3, \sec\alpha = -\frac{3}{2\sqrt{2}} = -\frac{3\sqrt{2}}{4},$$
$$\tan\alpha = \frac{-1/3}{-2\sqrt{2}/3} = \frac{\sqrt{2}}{4},$$
and $\cot\alpha = \frac{4}{\sqrt{2}} = 2\sqrt{2}.$

22. Since α is in Quadrant II, we get
$$\sin\alpha = \sqrt{1 - \left(-\frac{1}{4}\right)^2} = \frac{\sqrt{15}}{4},$$
$$\sec\alpha = -4, \csc\alpha = \frac{4}{\sqrt{15}} = \frac{4\sqrt{15}}{15},$$

$$\tan\alpha = \frac{\sqrt{15}/4}{-1/4} = -\sqrt{15},$$
and $\cot\alpha = -\frac{1}{\sqrt{15}} = -\frac{\sqrt{15}}{15}.$

23. Since α is in Quadrant I, we get
$$\sec\alpha = \sqrt{1 + \left(\frac{3}{5}\right)^2} = \frac{\sqrt{34}}{5},$$
$$\cot\alpha = \frac{5}{3}, \cos\alpha = \frac{5}{\sqrt{34}} = \frac{5\sqrt{34}}{34},$$
$$\csc\alpha = \sqrt{1 + \left(\frac{5}{3}\right)^2} = \frac{\sqrt{34}}{3},$$
and $\sin\alpha = \frac{3}{\sqrt{34}} = \frac{3\sqrt{34}}{34}.$

24. Since α is in Quadrant IV, we get
$$\cos\alpha = \sqrt{1 - \left(-\frac{4}{5}\right)^2} = \frac{3}{5},$$
$$\csc\alpha = -\frac{5}{4}, \sec\alpha = \frac{5}{3},$$
$$\tan\alpha = \frac{-4/5}{3/5} = -\frac{4}{3}, \text{ and}$$
$$\cot\alpha = -\frac{3}{4}.$$

25. $r\omega$

26. $\sin\alpha = \dfrac{y}{r}, \cos\alpha = \dfrac{x}{r}$

27. $\sin\alpha = y, \cos\alpha = x$

28. fundamental

29. amplitude

30. phase shift

31. frequency

32. vertical asymptote

33. even

34. odd

For Thought

1. True, the sum of the measurements of the three angles is 180°.

2. False, since similar triangles have the same corresponding angles but their corresponding sides are not necessarily equal.

3. True, since three angles do not uniquely determine a triangle.

4. False, $a \sin 17° = 88 \sin 9°$ and $a = \dfrac{88 \sin 9°}{\sin 17°}$.

5. False, since $\alpha = \sin^{-1}\left(\dfrac{5 \sin 44°}{18}\right) \approx 11°$ and $\alpha = 180 - 11° = 169°$.

6. True, since $\sin \beta = \dfrac{2.3 \sin 39°}{1.6}$.

7. True, since $\dfrac{\sin 60°}{\sqrt{3}} = \dfrac{\sqrt{3}/2}{\sqrt{3}} = \dfrac{1}{2}$ and $\dfrac{\sin 30°}{1} = \sin 30° = \dfrac{1}{2}$.

8. False, a triangle exists since $a = 500$ is bigger than $h = 10 \sin 60° \approx 8.7$.

9. True, since the triangle that exists is a right triangle.

10. False, there exists only one triangle and it is an obtuse triangle.

5.1 Exercises

1. oblique

3. ambiguous

5. Note $\gamma = 180° - (64° + 72°) = 44°$.

By the sine law $\dfrac{b}{\sin 72°} = \dfrac{13.6}{\sin 64°}$ and $\dfrac{c}{\sin 44°} = \dfrac{13.6}{\sin 64°}$. Then

$$b = \dfrac{13.6}{\sin 64°} \cdot \sin 72° \approx 14.4$$

and

$$c = \dfrac{13.6}{\sin 64°} \cdot \sin 44° \approx 10.5.$$

7. Note $\beta = 180° - (12.2° + 33.6°) = 134.2°$.

By the sine law $\dfrac{a}{\sin 12.2°} = \dfrac{17.6}{\sin 134.2°}$ and $\dfrac{c}{\sin 33.6°} = \dfrac{17.6}{\sin 134.2°}$. Then

$$a = \dfrac{17.6}{\sin 134.2°} \cdot \sin 12.2° \approx 5.2$$

and

$$c = \dfrac{17.6}{\sin 134.2°} \cdot \sin 33.6° \approx 13.6.$$

9. Note $\beta = 180° - (10.3° + 143.7°) = 26°$.

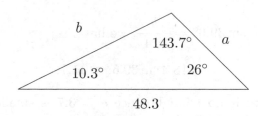

Since

$$\dfrac{a}{\sin 10.3°} = \dfrac{48.3}{\sin 143.7°}$$

and

$$\dfrac{b}{\sin 26°} = \dfrac{48.3}{\sin 143.7°}$$

we have

$$a = \dfrac{48.3}{\sin 143.7°} \cdot \sin 10.3° \approx 14.6$$

and

$$b = \dfrac{48.3}{\sin 143.7°} \cdot \sin 26° \approx 35.8$$

11. Note $\alpha = 180° - (120.7° + 13.6°) = 45.7°$.

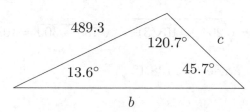

Since

$$\dfrac{c}{\sin 13.6°} = \dfrac{489.3}{\sin 45.7°}$$

and

$$\dfrac{b}{\sin 120.7°} = \dfrac{489.3}{\sin 45.7°}$$

we have

$$c = \frac{489.3}{\sin 45.7°} \cdot \sin 13.6° \approx 160.8 \text{ and}$$

$$b = \frac{489.3}{\sin 45.7°} \cdot \sin 120.7° \approx 587.9$$

13. Draw angle $\alpha = 39.6°$ and let h be the height.

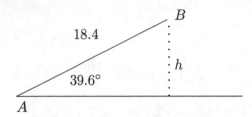

Since $\sin 39.6° = \dfrac{h}{18.4°}$, we have

$$h = 18.4 \sin 39.6° \approx 11.7.$$

There is no triangle since $a = 3.7$ is smaller than $h \approx 11.7$.

15. Draw angle $\gamma = 60°$ and let h be the height.

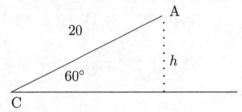

Since

$$h = 20 \sin 60° = 10\sqrt{3}$$

and $c = h$, there is exactly one triangle and it is a right triangle. Then $\beta = 90°$ and $\alpha = 30°$. By the Pythagorean Theorem,

$$a = \sqrt{20^2 - (10\sqrt{3})^2} = \sqrt{400 - 300} = 10.$$

17. Draw angle $\beta = 138.1°$.

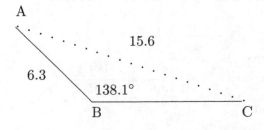

There is one triangle. Apply the sine law.

$$\frac{15.6}{\sin 138.1°} = \frac{6.3}{\sin \gamma}$$

$$\sin \gamma = \frac{6.3 \sin 138.1°}{15.6}$$

$$\sin \gamma \approx 0.2697$$

$$\gamma = \sin^{-1}(0.2697) \approx 15.6°$$

So $\alpha = 180° - (15.6° + 138.1°) = 26.3°$.

Using the sine law, we obtain

$$a = \frac{15.6}{\sin 138.1°} \sin 26.3° \approx 10.3.$$

19. Draw angle $\beta = 32.7°$ and let h be the height.

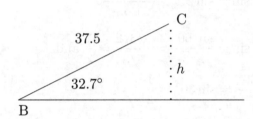

Since $h = 37.5 \sin 32.7° \approx 20.3$ and $20.3 < b < 37.5$, there are two triangles and they are given by

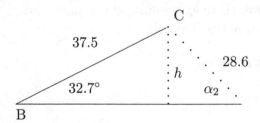

and

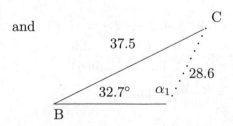

Apply the sine law to the acute triangle.

$$\frac{28.6}{\sin 32.7°} = \frac{37.5}{\sin \alpha_2}$$

$$\sin \alpha_2 = \frac{37.5 \sin 32.7°}{28.6}$$

$$\sin \alpha_2 \approx 0.708$$

$$\alpha_2 = \sin^{-1}(0.708) \approx 45.1°$$

So $\gamma_2 = 180° - (45.1° + 32.7°) = 102.2°$.

By the sine law,

$$c_2 = \frac{28.6}{\sin 32.7°} \sin 102.2° \approx 51.7.$$

On the obtuse triangle, we find
$\alpha_1 = 180° - \alpha_2 = 134.9°$ and
$\gamma_1 = 180° - (134.9° + 32.7°) = 12.4°$.

By the sine law,

$$c_1 = \frac{28.6}{\sin 32.7°} \sin 12.4° \approx 11.4.$$

21. Draw angle $\gamma = 99.6°$. Note, there is exactly one triangle since $12.4 > 10.3$.

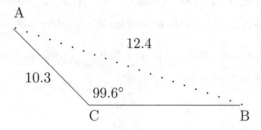

By the sine law, we obtain

$$\frac{12.4}{\sin 99.6°} = \frac{10.3}{\sin \beta}$$

$$\sin \beta = \frac{10.3 \sin 99.6°}{12.4}$$

$$\sin \beta \approx 0.819$$

$$\beta = \sin^{-1}(0.819) \approx 55.0°.$$

So $\alpha = 180° - (55.0° + 99.6°) = 25.4°$.

Using the sine law, we find

$$a = \frac{12.4}{\sin 99.6°} \sin 25.4° \approx 5.4.$$

23. Let x be the number of miles flown along I-20.

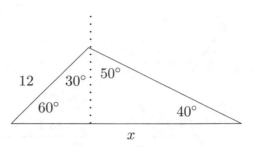

Applying the sine law, we obtain

$$\frac{x}{\sin 80°} = \frac{12}{\sin 40°}.$$

Then $x \approx 18.4$ miles.

25. Let x and y be the lengths of the missing sides.

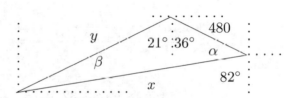

There is a 21° angle because of the $S21°W$ direction. There are 36° and 82° angles because opposite angles are equal and because of the directions $N36°W$ and $N82°E$.
Note $\alpha = 180° - (82° + 36°) = 62°$ and $\beta = 180 - (21° + 36° + 62°) = 61°$.

By the sine law, we find

$$x = \frac{480}{\sin 61°} \sin 57° \approx 460.27$$

and

$$y = \frac{480}{\sin 61°} \sin 62° \approx 484.57.$$

The perimeter is $x + y + 480 \approx 1425$ ft.

27. Applying the sine law, we find

$$x = \frac{19.2 \sin 82°}{\sin 30°} \approx 38.0 \text{ ft.}$$

29. Let h be the height of the tower.

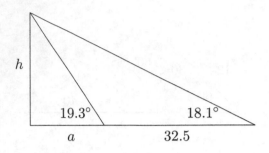

Using right triangle trigonometry, we get

$$\tan 19.3° = \frac{h}{a}$$

or

$$a = \frac{h}{\tan 19.3°}.$$

Similarly, we have

$$\tan 18.1° = \frac{h}{a + 32.5}.$$

Then

$$\tan 18.1°(a + 32.5) = h$$
$$a\tan 18.1° + 32.5\tan 18.1° = h$$
$$\frac{h}{\tan 19.3°} \cdot \tan 18.1° + 32.5\tan 18.1° = h$$
$$h \cdot \frac{\tan 18.1°}{\tan 19.3°} + 32.5\tan 18.1° = h.$$

Solving for h, we find that the height of the tower is

$$h \approx 159.4 \text{ ft.}$$

31. Note, $\tan\gamma = 6/12$ and $\gamma = \tan^{-1}(0.5) \approx$ 26.565°. Also, $\tan\alpha = 3/12$ and $\alpha = \tan^{-1}(0.25) \approx 14.036°$.

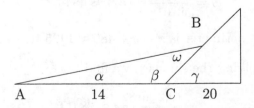

The remaining angles are $\beta = 153.435°$ and $\omega = 12.529°$. By the sine law, we obtain

$$\frac{AB}{\sin 153.435°} = \frac{14}{\sin 12.529°}$$

and

$$\frac{BC}{\sin 14.036°} = \frac{14}{\sin 12.529°}.$$

Then $AB \approx 28.9$ ft and $BC \approx 15.7$ ft.

33. By the sine law, we get

$$x = \frac{24\sin 47°}{\sin 104°} \approx 18.1 \text{ in.}$$

35. Let t be the number of seconds since the cruise missile was spotted.

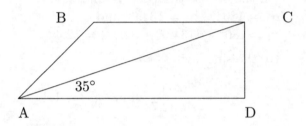

Let β be the angle at B. The angle formed by BAC is $180° - 35° - \beta$. After t seconds, the cruise missile would have traveled $548\dfrac{t}{3600}$ miles and the projectile $688\dfrac{t}{3600}$ miles. Using the law of sines, we have

$$\frac{\dfrac{548t}{3600}}{\sin(145° - \beta)} = \frac{\dfrac{688t}{3600}}{\sin 35°}$$

$$\frac{548}{\sin(145° - \beta)} = \frac{688}{\sin 35°}$$

$$\beta = 145° - \sin^{-1}\left(\frac{548\sin 35°}{688}\right)$$

$$\beta \approx 117.8°.$$

Then angle BAC is 27.2°. The angle of elevation of the projectile must be angle DAB which is 62.2° $(= 35° + 27.2°)$.

37. Let t be the number of seconds it takes the fox to catch the rabbit. The distances travelled by the fox and rabbit are indicated below.

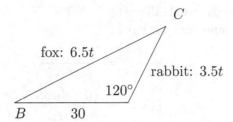

Apply the sine law as follows:

$$\frac{6.5t}{\sin 120°} = \frac{3.5t}{\sin B}$$

$$\sin B = \frac{3.5\sqrt{3}}{13}$$

Note, $C = 60° - \arcsin\left(\frac{3.5\sqrt{3}}{13}\right)$. Then

$$\frac{30}{\sin C} = \frac{3.5t}{\sin B}$$

$$t = \frac{30 \sin B}{3.5 \sin C} = 7.5$$

It will take 7.5 sec to catch the rabbit.

41. a) 1   **b)** $\dfrac{3\pi}{4}$ **c)** $-\sqrt{3}$

    d) $-\dfrac{2\sqrt{3}}{3}$ **e)** $-\sqrt{2}$ **f)** $-\dfrac{\pi}{6}$

43. a) $\dfrac{2\pi}{\pi} = 2$ **b)** $\dfrac{2\pi}{3}$

    c) $\dfrac{\pi}{2\pi} = \dfrac{1}{2}$ **d)** $\dfrac{2\pi}{2} = \pi$

45. a) Odd   **b)** Even

    c) Even   **d)** Even

47. Triangle $\triangle ABC$ is an isosceles triangle. Also, $\triangle ACD$ and $\triangle ABD$ are isosceles triangles. Then $AC = AD = 2$. We apply the Angle Bisector Theorem in Exercise 50.

$$\frac{CD+2}{2} = \frac{2}{CD}.$$

Solving for CD, we find $CD = \sqrt{5} - 1$.

For Thought

1. True, since $\cos 90° = 0$ in the law of cosines.

2. False, $a = \sqrt{c^2 + b^2 - 2bc \cos \alpha}$.

3. False, $c^2 = a^2 + b^2 - 2ab \cos \gamma$.

4. True, this follows from the sine law.

5. False, it has only one solution in $[0°, 180°]$.

6. True, since the sum of the angles is $180°$.

7. True, since $\beta = \sin^{-1}(0.1235)$ or $\beta = 180° - \sin^{-1}(0.1235)$.

8. True, since the law of cosines will be used and cosine is a one-to-one function in $[0°, 180°]$.

9. True, since $\cos \gamma = \dfrac{3.4^2 + 4.2^2 - 8.1^2}{2(3.4)(4.2)} \approx -1.27$ has no real solution γ.

10. False, there is exactly one triangle.

5.2 Exercises

1. law of cosines

3. cosines

5. By the cosine law, we obtain

$$c = \sqrt{3.1^2 + 2.9^2 - 2(3.1)(2.9) \cos 121.3°}$$

$$\approx 5.23 \approx 5.2. \text{ By the sine law, we find}$$

$$\frac{3.1}{\sin \alpha} = \frac{5.23}{\sin 121.3°}$$

$$\sin \alpha = \frac{3.1 \sin 121.3°}{5.23}$$

$$\sin \alpha \approx 0.50647$$

$$\alpha \approx \sin^{-1}(0.50647) \approx 30.4°.$$

Then $\beta = 180° - (30.4° + 121.3°) = 28.3°$.

7. By the cosine law, we find

$$\cos \beta = \frac{6.1^2 + 5.2^2 - 10.3^2}{2(6.1)(5.2)} \approx -0.6595$$

and so

$$\beta \approx \cos^{-1}(-0.6595) \approx 131.3°.$$

By the sine law,

$$\frac{6.1}{\sin\alpha} = \frac{10.3}{\sin 131.3°}$$

$$\sin\alpha = \frac{6.1\sin 131.3°}{10.3}$$

$$\sin\alpha \approx 0.4449$$

$$\alpha \approx \sin^{-1}(0.4449) \approx 26.4°.$$

So $\gamma = 180° - (26.4° + 131.3°) = 22.3°$.

9. By the cosine law,

$$b = \sqrt{2.4^2 + 6.8^2 - 2(2.4)(6.8)\cos 10.5°}$$

$$\approx 4.46167 \approx 4.5 \text{ and}$$

$$\cos\alpha = \frac{2.4^2 + 4.46167^2 - 6.8^2}{2(2.4)(4.46167)} \approx -0.96066.$$

So $\alpha = \cos^{-1}(-0.96066) \approx 163.9°$ and
$\gamma = 180° - (163.9° + 10.5°) = 5.6°$

11. By the cosine law,

$$\cos\alpha = \frac{12.2^2 + 8.1^2 - 18.5^2}{2(12.2)(8.1)} \approx -0.6466.$$

Then $\alpha = \cos^{-1}(-0.6466) \approx 130.3°$.
By the sine law,

$$\frac{12.2}{\sin\beta} = \frac{18.5}{\sin 130.3°}$$

$$\sin\beta = \frac{12.2\sin 130.3°}{18.5}$$

$$\sin\beta \approx 0.5029$$

$$\beta \approx \sin^{-1}(0.5029) \approx 30.2°$$

So $\gamma = 180° - (30.2° + 130.3°) = 19.5°$.

13. By the cosine law, we obtain

$$a = \sqrt{9.3^2 + 12.2^2 - 2(9.3)(12.2)\cos 30°}$$

$$\approx 6.23 \approx 6.2 \text{ and}$$

$$\cos\gamma = \frac{6.23^2 + 9.3^2 - 12.2^2}{2(6.23)(9.3)} \approx -0.203.$$

So $\gamma = \cos^{-1}(-0.203) \approx 101.7°$ and
$\beta = 180° - (101.7° + 30°) = 48.3°$.

15. By the cosine law,

$$\cos\beta = \frac{6.3^2 + 6.8^2 - 7.1^2}{2(6.3)(6.8)} \approx 0.4146.$$

So $\beta = \cos^{-1}(0.4146) \approx 65.5°$.
By the sine law, we have

$$\frac{6.8}{\sin\gamma} = \frac{7.1}{\sin 65.5°}$$

$$\sin\gamma = \frac{6.8\sin 65.5°}{7.1}$$

$$\sin\gamma \approx 0.8715$$

$$\gamma \approx \sin^{-1}(0.8715) \approx 60.6°.$$

So $\alpha = 180° - (60.6° + 65.5°) = 53.9°$.

17. Note, $\alpha = 180° - 25° - 35° = 120°$.
Then by the sine law, we obtain

$$\frac{7.2}{\sin 120°} = \frac{b}{\sin 25°} = \frac{c}{\sin 35°}$$

from which we have

$$b = \frac{7.2\sin 25°}{\sin 120°} \approx 3.5$$

and

$$c = \frac{7.2\sin 35°}{\sin 120°} \approx 4.8.$$

19. There is no such triangle. Note, $a + b = c$ and in a triangle the sum of the lengths of two sides is greater than the length of the third side.

21. One triangle exists. The angles are uniquely determined by the law of cosines.

23. There is no such triangle since the sum of the angles in a triangle is $180°$.

25. Exactly one triangle exists. This is seen by constructing a $179°$-angle with two sides that have lengths 1 and 10. The third side is constructed by joining the endpoints of the first two sides.

27. Consider the figure below.

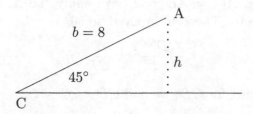

Note, $h = 8\sin 45° = 4\sqrt{2}$. So the minimum value of c so that we will be able to make a triangle is $4\sqrt{2}$. Since $c = 2$, no such triangle is possible.

29. Recall, a central angle α in a circle of radius r intercepts a chord of length $r\sqrt{2 - 2\cos\alpha}$. Since $r = 30$ and $\alpha = 19°$, the length is

$$30\sqrt{2 - 2\cos 19°} \approx 9.90 \ \text{ft}.$$

31. Note, a central angle α in a circle of radius r intercepts a chord of length $r\sqrt{2 - 2\cos\alpha}$. Since

$$921 = r\sqrt{2 - 2\cos 72°}$$

(where $360 \div 5 = 72$), we obtain

$$r = \frac{921}{\sqrt{2 - 2\cos 72°}} \approx 783.45 \ \text{ft}.$$

33. After 6 hours, Jan hiked a distance of 24 miles and Dean hiked 30 miles. Let x be the distance between them after 6 hrs.

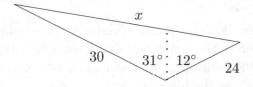

By the cosine law, we find
$$x = \sqrt{30^2 + 24^2 - 2(30)(24)\cos 43°} = \sqrt{1476 - 1440\cos 43°} \approx 20.6 \ \text{miles}.$$

35. By the cosine law, we find

$$\begin{aligned}
\cos\alpha &= \frac{1.2^2 + 1.2^2 - 0.4^2}{2(1.2)(1.2)} \\
\cos\alpha &\approx 0.9444 \\
\alpha &\approx \cos^{-1}(0.9444) \\
\alpha &\approx 19.2°.
\end{aligned}$$

37. Let α, β, and γ be the angles at pipes A, B, and C. The length of the sides of the triangle are 5, 6, and 7. By the cosine law,

$$\begin{aligned}
\cos\alpha &= \frac{5^2 + 6^2 - 7^2}{2(5)(6)} \\
\cos\alpha &= 0.2 \\
\alpha &= \cos^{-1}(0.2) \\
\alpha &\approx 78.5°.
\end{aligned}$$

By the sine law,

$$\begin{aligned}
\frac{6}{\sin\beta} &= \frac{7}{\sin 78.5°} \\
\sin\beta &\approx 0.8399 \\
\beta &\approx \sin^{-1}(0.8399) \\
\beta &\approx 57.1°.
\end{aligned}$$

Then $\gamma = 180° - (57.1° + 78.5°) = 44.4°$.

39. By the cosine law,
$$AB = \sqrt{5.3^2 + 7.6^2 - 2(5.3)(7.6)\cos 28°} = \sqrt{85.85 - 80.56\cos 28°} \approx 3.8 \ \text{miles}.$$

By using the exact value of AB, we get

$$\begin{aligned}
\cos(\angle CBA) &= \frac{AB^2 + 5.3^2 - 7.6^2}{2(AB)(5.3)} \\
\angle CBA &= \cos^{-1}\left(\frac{AB^2 + 5.3^2 - 7.6^2}{2(AB)(5.3)}\right) \\
\angle CBA &\approx 111.6°
\end{aligned}$$

and $\angle CAB = 180° - (111.6° + 28°) = 40.4°$.

41. The pentagon consists of 5 chords each of which intercepts a $\dfrac{360°}{5} = 72°$ angle.
By the cosine law, the length of a chord is given by

$$\sqrt{10^2 + 10^2 - 2(10)(10)\cos 72°} = $$

$$\sqrt{200 - 200\cos 72°} \approx 11.76 \ \text{m}.$$

43. The lower-left corner is the origin $(0,0)$.

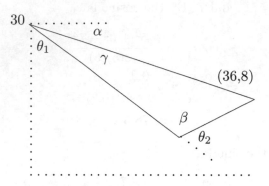

Note $\tan\alpha = 22/36$ and

$$\alpha = \tan^{-1}(22/36) \approx 31.4°.$$

The distance between $(36,8)$ and $(0,30)$ is approximately 42.19. By the cosine law,

$$\beta = \cos^{-1}\left(\frac{30^2 + 30^2 - 42.19^2}{2(30)(30)}\right)$$

$$\beta \approx 89.4°.$$

So $\theta_2 = 180° - 89.4° = 90.6°$.

By the sine law, we find

$$\frac{30}{\sin\gamma} = \frac{42.19}{\sin 89.4°}$$

$$\gamma = \sin^{-1}\left(\frac{30\sin 89.4°}{42.19}\right) \approx 45.3°.$$

Then $\theta_1 = 90° - (45.3° + 31.4°) = 13.3°$.

45. a) Let α_m and α_M be the minimum and maximum values of α, respectively. By the law of cosines, we get

$$865,000^2 = 2(91,400,000)^2 - 2(91,400,000)^2\cos\alpha_M.$$

Then

$$\alpha_M = \cos^{-1}\left(\frac{2(91400000)^2 - 865000^2}{2(91400000)^2}\right)$$

$$\alpha_M \approx 0.54°.$$

Likewise,

$$\alpha_m = \cos^{-1}\left(\frac{2(94500000)^2 - 865000^2}{2(94500000)^2}\right)$$

$$\alpha_m \approx 0.52°.$$

b) Let β_m and β_M be the minimum and maximum values of β, respectively. By the law of cosines, one obtains

$$2163^2 = 2(225,800)^2 - 2(225,800)^2\cos\beta_M.$$

Then

$$\beta_M = \cos^{-1}\left(\frac{2(225800)^2 - 2163^2}{2(225800)^2}\right)$$

$$\beta_M \approx 0.55°.$$

Likewise,

$$\beta_m = \cos^{-1}\left(\frac{2(252000)^2 - 2163^2}{2(252000)^2}\right)$$

$$\beta_m \approx 0.49°.$$

c) Yes, even in perfect alignment a total eclipse may not occur, for instance when $\beta = 0.49°$ and $\alpha = 0.52°$.

47. Let d_b and d_h be the distance from the bear and hiker, respectively, to the base of the tower. Then $d_b = 150\tan 80°$ and $d_h = 150\tan 75°$.

Since the line segments joining the base of the tower to the bear and hiker form a 45° angle, by the cosine law the distance, d, between the bear and the hiker is

$$d = \sqrt{d_b^2 + d_h^2 - 2(d_b)(d_h)\cos 45°}$$

$$\approx \big((850.69)^2 + (559.81)^2 -$$

$$2(850.69)(559.81)\cos 45°\big)^{1/2}$$

$$\approx 603 \text{ feet.}$$

49. Using the cosine law, we obtain

$$a = \sqrt{2r^2 - 2r^2\cos(\theta)} = \sqrt{4r^2\frac{1-\cos\theta}{2}} = 2r\sin(\theta/2).$$

51. Note, $\gamma = 180° - 108.1° - 18.6° = 53.3°$. By the sine law, we obtain

$$a = \frac{28.6\sin 108.1°}{\sin 53.3°} \approx 33.9$$

$$b = \frac{28.6\sin 18.6°}{\sin 53.3°} \approx 11.4$$

53. Since the y-values of the key points are 3 ± 2, we find $A = 2$ and $D = 3$. Since the first key point is $(\pi/4, 3)$, the phase shift is $C = \pi/4$. Since the difference between the first and last y-values is the period, we find

$$\frac{2\pi}{B} = \frac{5\pi}{4} - \frac{\pi}{4} = \pi$$

and $B = 2$. The equation is

$$y = 2\sin\left(2\left(x - \frac{\pi}{4}\right)\right) + 3$$

55. a) $270° \cdot \dfrac{\pi}{180°} = \dfrac{3\pi}{2}$ **b)** $315° \cdot \dfrac{\pi}{180°} = \dfrac{7\pi}{4}$

c) $-210° \cdot \dfrac{\pi}{180°} = -\dfrac{7\pi}{6}$ **d)** $120° \cdot \dfrac{\pi}{180°} = \dfrac{2\pi}{3}$

57. When a point on a circle with radius r is rotated through an angle of $\pi/2$, the distance the point rotates is

$$s = r\frac{\pi}{2}.$$

The sum of the distances traveled by point A is

$$\sqrt{45}\frac{\pi}{2} + 3\frac{\pi}{2} + 0 + 6\frac{\pi}{2} = \frac{(3\sqrt{5} + 9)\pi}{2} \ \ \text{ft.}$$

For Thought

1. False, rather in a right triangle the area is one-half the product of its legs.

2. True

3. False, rather the area is one-half the product of two lengths of two sides and the sine of the included angle.

4. True

5. True, since one can use Heron's formula.

5.3 Exercises

1. $bh/2$

3. Since two sides and an included angle are given, the area is

$$A = \frac{1}{2}(12.9)(6.4)\sin 13.7° \approx 9.8.$$

5. Draw angle $\alpha = 39.4°$.

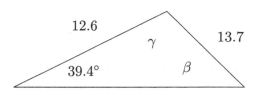

By the sine law, we obtain

$$\frac{12.6}{\sin\beta} = \frac{13.7}{\sin 39.4°}$$

$$\sin\beta = \frac{12.6\sin 39.4°}{13.7}$$

$$\beta = \sin^{-1}\left(\frac{12.6\sin 39.4°}{13.7}\right)$$

$$\beta \approx 35.7°.$$

Then $\gamma = 180° - (35.7° + 39.4°) = 104.9°$.

The area is $A = \dfrac{1}{2} \cdot ab\sin\gamma = $

$$\frac{1}{2} \cdot (13.7)(12.6)\sin 104.9° \approx 83.4.$$

7. Draw angle $\alpha = 42.3°$.

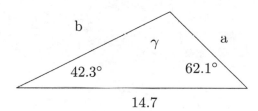

Note $\gamma = 180° - (42.3° + 62.1°) = 75.6°$.
By the sine law,

$$\frac{b}{\sin 62.1°} = \frac{14.7}{\sin 75.6°}$$

$$b = \frac{14.7}{\sin 75.6°} \cdot \sin 62.1°$$

$$b \approx 13.41.$$

The area is $A = \dfrac{1}{2}bc\sin\alpha = $

$$\frac{1}{2}(13.41)(14.7)\sin 42.3° \approx 66.3.$$

9. Draw angle $\alpha = 56.3°$.

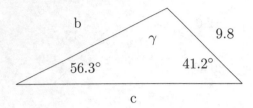

Note $\gamma = 180° - (56.3° + 41.2°) = 82.5°$.
By the sine law, we obtain

$$\frac{c}{\sin 82.5°} = \frac{9.8}{\sin 56.3°}$$

$$c = \frac{9.8}{\sin 56.3°}\sin 82.5°$$

$$c \approx 11.679.$$

The area is $A = \dfrac{1}{2}ac\sin\beta =$

$$\frac{1}{2}(9.8)(11.679)\sin 41.2° \approx 37.7.$$

11. Note, the area of the triangle below is

$$\frac{1}{2}\cdot(1.5)(1.5\sqrt{3}) = 1.125\sqrt{3}.$$

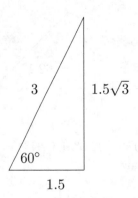

The trapezoid in the problem can be divided into two triangles and a rectangle with dimensions 2.7 by $1.5\sqrt{3}$. Thus, the area of the trapezoid is the area of the rectangle plus twice the area of the triangle shown above. That is, the area of the trapezoid is

$$2.7(1.5\sqrt{3}) + 2(1.125\sqrt{3}) = 6.3\sqrt{3} \approx 11 \text{ ft}^2.$$

13. Divide the given 4-sided polygon into two triangles by drawing the diagonal that

connects the 60° angle to the 135° angle. On each triangle two sides and an included angle are given. The area of the polygon is equal to the sum of the areas of the two triangles. Namely,

$$\frac{1}{2}(4)(10)\sin 120° + \frac{1}{2}(12+2\sqrt{3})(2\sqrt{6})\sin 45° =$$

$$20(\sqrt{3}/2) + \frac{1}{2}(24\sqrt{6} + 4\sqrt{18})(\sqrt{2}/2) =$$

$$10\sqrt{3} + \frac{1}{2}(12\sqrt{12} + 2\sqrt{36}) =$$

$$10\sqrt{3} + 6\sqrt{12} + \sqrt{36} = 10\sqrt{3} + 12\sqrt{3} + 6 \approx$$

$$44 \text{ square miles.}$$

15. Note,

$$S = \frac{16 + 9 + 10}{2} = 17.5.$$

The area is

$$A = \sqrt{17.5(17.5 - 16)(17.5 - 9)(17.5 - 10)}$$

$$= \sqrt{17.5(1.5)(8.5)(7.5)} \approx 40.9.$$

17. Note,

$$S = \frac{3.6 + 9.8 + 8.1}{2} = 10.75.$$

The area is

$$\sqrt{10.75(10.75 - 3.6)(10.75 - 9.8)(10.75 - 8.1)}$$

$$= \sqrt{10.75(7.15)(0.95)(2.65)} \approx 13.9.$$

19. Note,

$$S = \frac{346 + 234 + 422}{2} = 501.$$

The area is

$$\sqrt{501(501 - 346)(501 - 234)(501 - 422)} =$$

$$\sqrt{501(155)(267)(79)} \approx 40,471.9.$$

21. Since the base is 20 and the height is 10, the area is $\dfrac{1}{2}bh = \dfrac{1}{2}(20)(10) = 100$.

23. Since two sides and an included angle are given, the area is

$$\frac{1}{2}(6)(8)\sin 60° \approx 20.8.$$

25. Note, $S = \dfrac{9 + 5 + 12}{2} = 13$.

The area is $\sqrt{13(13 - 9)(13 - 5)(13 - 12)} =$

$$\sqrt{13(4)(8)(1)} \approx 20.4.$$

27. The kite consists of two equal triangles. The area of the kite is twice the area of the triangle.

Then the area of the kite is

$$2\left(\frac{1}{2}\right)(24)(18)\sin 40° \approx 277.7 \text{ in.}^2.$$

29. The largest angle γ is opposite the 13-inch side. By the cosine law, we find

$$\gamma = \cos^{-1}\left(\frac{8^2 + 9^2 - 13^2}{2(8)(9)}\right) \approx 99.6°.$$

Thus, the area is

$$\frac{1}{2}(8)(9)\sin(99.6°) \approx 35.5 \text{ in.}^2.$$

31. Since the area is $A = \dfrac{1}{2}ab\sin\gamma$, the area is maximized when $\gamma = 90°$. Then the maximum area is $A = \dfrac{1}{2}(2)(2) = 2$ square ft.

33. a) The area of triangle is one-half the product of two sides and the sine of the included angle. If the sides have both length r, and the central angle is α, then the area of the triangle is

$$A_T = \frac{1}{2}r^2\sin\alpha.$$

b) The area of a sector is proportional to the area of a circle. If the central angle is α and the radius is r, the area of the sector is

$$A_s = \frac{r^2\alpha}{2}.$$

c) The area A_L of a lens-shaped region is the difference of the area of a sector and the area of a triangle, see parts a) and b). Then

$$A_L = \frac{r^2\alpha}{2} - \frac{1}{2}r^2\sin\alpha = \frac{r^2}{2}(\alpha - \sin\alpha).$$

35. Let x be the length of the third side.

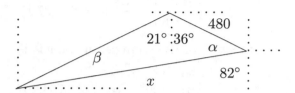

There is a 21° angle because of the $S21°W$ direction. There are 36° and 82° angles because opposite angles are equal and because of the directions $N36°W$ and $N82°E$. Note,

$$\alpha = 180° - (82° + 36°) = 62°$$

and $\beta = 180 - (21° + 36° + 62°) = 61°$. By the sine law, we obtain

$$x = \frac{480}{\sin 61°}\sin 57°.$$

The area is

$$\frac{1}{2}(480x)\sin 62° \approx 97,534.8 \text{ sq ft.}$$

37. Note the angles in the quadrilateral property.

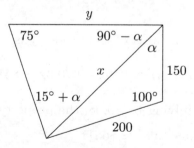

By the cosine law, we obtain

$$x = \sqrt{200^2 + 150^2 - 2(200)(150)\cos 100°}.$$

Then the area of the triangle on the right

$$A_R = \frac{1}{2}200(150)\sin 100° \approx 14,772.1163.$$

By the sine law, we find that in degrees

$$\alpha = \sin^{-1}\left(\frac{200\sin 100°}{x}\right) \approx 46.8355°.$$

Similarly, by the sine law, we get

$$y = \frac{x\sin(15° + \alpha)}{\sin 75°} \approx 246.4597$$

and the area of the triangle on the left is

$$A_L = \frac{1}{2}(xy)\sin(90° - \alpha) \approx 22,764.2076.$$

Thus, the area of the property is

$$A_R + A_L \approx 37,536.3 \text{ ft}^2.$$

39. Consider the lens-shaped region whose arc length is $s = 88.1$ ft. Since $s = r\alpha = 80\alpha$, the central angle is $\alpha = 88.1/80$.

The area A_L of the lens-shaped region is

$$A_L = \frac{80^2}{2}(\alpha - \sin \alpha) \approx 670.32$$

by Exercise 33c.

Join a line segment through the vertices of the lot that lie on the circle. This segment together with the other three sides of the lot form a trapezoid. The area A_T of the trapezoid is

$$A_T = \frac{80}{2}(102.5 + 127.1) = 9184.$$

Then the area of the property is

$$A_T - A_L \approx 9184 - 670.32 \approx 8513.68 \text{ ft}^2.$$

Multiplying by \$0.08, the property tax is $8513.68(0.08) \approx \$681$.

43. Note,

$$S = \frac{37 + 48 + 86}{2} = 85.5.$$

By Heron's formula, the area of the triangle is suppose to be

$$\sqrt{85.5(85.5 - 37)(85.5 - 48)(85.5 - 86)}.$$

But this area is undefined since we have the square root of a negative number. Thus, no triangle exists with sides 37, 48, and 86.

45. Let $a = 6$, $b = 9$, and $c = 13$. Then

$$4b^2c^2 = 54,756$$

and

$$(b^2 + c^2 - a^2)^2 = 45,796.$$

The area is given by

$$\frac{1}{4}\sqrt{4b^2c^2 - (b^2 + c^2 - a^2)^2} =$$

$$\frac{1}{4}\sqrt{54,756 - 45,796} = \frac{1}{4}\sqrt{8960} =$$

$$4\sqrt{35} \approx 23.7 \text{ ft}^2.$$

Next, we verify that that

$$A_1 = \frac{1}{4}\sqrt{4b^2c^2 - (b^2 + c^2 - a^2)^2}$$

or equivalently

$$\sqrt{\frac{4b^2c^2 - (b^2 + c^2 - a^2)^2}{16}}$$

gives the area of a triangle. To do this, we will use Heron's formula. Let $s = \dfrac{a+b+c}{2}$. Since it can be shown that

$$s(s-a)(s-b)(s-c) = \frac{4b^2c^2 - (b^2 + c^2 - a^2)^2}{16}$$

it then follows that formula A_1 gives the area of a triangle.

49. Angle α is an acute angle. By the sine law,

$$\alpha = \sin^{-1}\left(\frac{19.4\sin 122.1°}{22.6}\right)$$

$$\alpha \approx 46.7°$$

Then $\gamma = 180° - 122.1° - \alpha \approx 11.2°$. Applying the sine law, we find

$$c = \frac{22.6\sin\gamma}{\sin 122.1°} \approx 5.2$$

51. Draw angle $\alpha = 33.2°$.

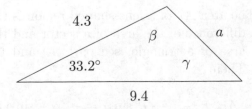

By the cosine law, we find

$$a^2 = 9.4^2 + 4.3^2 - 2(9.4)(4.3)\cos 33.2°$$
$$a \approx 6.3$$

Note, γ is an acute angle. By the sine law, we obtain

$$\gamma = \sin^{-1}\left(\frac{4.3\sin 33.2°}{a}\right) \approx 22.1°.$$

Then $\beta = 180° - 33.2° - \gamma \approx 124.7°$.

53. Observe $\sin 3x = \pm\dfrac{\sqrt{3}}{2}$. Then

$$3x = \frac{\pi}{3} + k\pi \text{ or } 3x = \frac{2\pi}{3} + k\pi$$

where k is an integer. Solve for x as follows:

$$x = \frac{\pi}{9} + \frac{3k\pi}{9} \text{ or } x = \frac{2\pi}{9} + \frac{3k\pi}{9}$$

If $k = 0, 1, 2$, the solutions are

$$x = \frac{\pi}{9}, \frac{2\pi}{9}, \frac{4\pi}{9}, \frac{5\pi}{9}, \frac{7\pi}{9}, \frac{8\pi}{9}$$

55. Using Heron's formula, the area of the triangle is

$$A = \sqrt{15(6)4)(5)}.$$

Let α, β, and γ be the angles included by sides 9 & 10, 9 & 11, and 10 & 11, respectively. By the cosine law, we find

$$\alpha = \cos^{-1}\left(\frac{9^2 + 10^2 - 11^2}{2(9)(10)}\right)$$

$$\beta = \cos^{-1}\left(\frac{9^2 + 11^2 - 10^2}{2(9)(11)}\right)$$

$$\gamma = \cos^{-1}\left(\frac{10^2 + 11^2 - 9^2}{2(10)(11)}\right)$$

Draw a sector with central angle α and radius 4, and the area of this sector is

$$S_1 = \frac{1}{2}\left(4^2\alpha\right) \approx 9.847675339.$$

Similarly, let S_2 and S_3 be the areas of the sectors with central angle β and radius 5, and central angle γ and radius 6, respectively.

Thus, the area that is not sprayed by any of the three sprinklers is

$$A - (S_1 + S_2 + S_3) \approx 3.850 \text{ meters}^2.$$

For Thought

1. True, since if $v = \langle x, y\rangle$ then $2v = \langle 2x, 2y\rangle$ and $\tan^{-1}\left(\dfrac{2y}{2x}\right) = \tan^{-1}\left(\dfrac{y}{x}\right)$ and $|2v| = \sqrt{4x^2 + 4y^2} = 2\sqrt{x^2 + y^2} = 2|v|$.

2. False, if $\boldsymbol{A} = \langle 1, 0\rangle$ and $\boldsymbol{B} = \langle 0, 1\rangle$ then $|\boldsymbol{A} + \boldsymbol{B}| = |\langle 1, 1\rangle| = \sqrt{2}$ and $|\boldsymbol{A}| + |\boldsymbol{B}| = 2$.

3. True, since if $A = \langle x, y\rangle$ then $-A = \langle -x, -y\rangle$ and $|-A| = \sqrt{(-x)^2 + (-y)^2} = \sqrt{x^2 + y^2} = |A|$.

4. True, since $\langle x, y\rangle + \langle -x, -y\rangle = \langle 0, 0\rangle$.

5. False, rather the parallelogram law says that the magnitude of $\boldsymbol{A} + \boldsymbol{B}$ is the length of a diagonal of the parallelogram formed by $\boldsymbol{A}$ and $\boldsymbol{B}$.

6. False, the direction angle is formed with the positive x-axis.

7. True, this follows from the fact that the horizontal component makes a $0°$-angle with the positive x-axis and $\cos\theta = $ adjacent/hypotenuse.

8. True, since $|\langle 3, -4\rangle| = \sqrt{3^2 + (-4)^2} = \sqrt{9 + 16} = 5$.

9. True, the direction angle of a vector is unchanged when it is multipied by a positive scalar.

10. True, since $r = \sqrt{(-2)^2 + 2^2} = \sqrt{8}$ and $\cos\theta = \dfrac{x}{r} = \dfrac{-2}{\sqrt{8}}$.

5.4 Exercises

1. vector

3. magnitude

5. parallelogram law

7. component

9. $A + B = 5\,j + 4\,i = 4\,i + 5\,j$

and

$$A - B = 5\,j - 4\,i = -4\,i + 5\,j$$

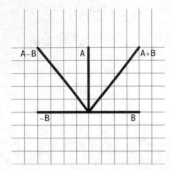

11. $A + B = (\,i + 3\,j\,) + (4\,i + j\,) = 5\,i + 4\,j$

and

$$A - B = (\,i + 3\,j\,) - (4\,i + j\,) = -3\,i + 2\,j$$

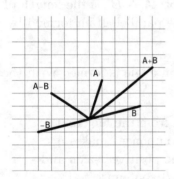

13. $A + B = (-\,i + 4\,j\,) + (4\,i\,) = 3\,i + 4\,j$

and

$$A - B = (-\,i + 4\,j\,) - (4\,i\,) = -5\,i + 4\,j$$

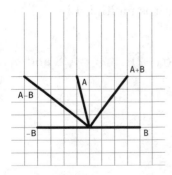

15. D   17.   E

19. B

21. $|\,v_x\,| = |4.5\cos 65.2°| = 1.9,$
$|\,v_y\,| = |4.5\sin 65.2°| = 4.1$

23. $|\,v_x\,| = |8000\cos 155.1°| \approx 7256.4,$
$|\,v_y\,| = |8000\sin 155.1°| \approx 3368.3$

25. $|\,v_x\,| = |234\cos 248°| \approx 87.7,$
$|\,v_y\,| = |234\sin 248°| \approx 217.0$

27. The magnitude is $\sqrt{\sqrt{3}^2 + 1^2} = 2.$
Since $\tan\alpha = 1/\sqrt{3}$, the direction angle is $\alpha = 30°$.

29. The magnitude is $\sqrt{(-\sqrt{2})^2 + \sqrt{2}^2} = 2.$
Since $\tan\alpha = -\sqrt{2}/\sqrt{2} = -1$, the direction angle is $\alpha = 135°$.

31. The magnitude is $\sqrt{8^2 + (-8\sqrt{3})^2} = 16.$
Since $\tan\alpha = -8\sqrt{3}/8 = -\sqrt{3}$, the direction angle is $\alpha = 300°$.

33. The magnitude is $\sqrt{5^2 + 0^2} = 5.$
Since the terminal point is on the positive x-axis, the direction angle is $0°$.

35. The magnitude is $\sqrt{(-3)^2 + 2^2} = \sqrt{13}.$
Since $\tan^{-1}(-2/3) \approx -33.7°$, the direction angle is $180° - 33.7° = 146.3°$.

37. The magnitude is $\sqrt{3^2 + (-1)^2} = \sqrt{10}.$
Since $\tan^{-1}(-1/3) \approx -18.4°$, the direction angle is $360° - 18.4° = 341.6°$.

39. $\langle 8\cos 45°, 8\sin 45° \rangle = \langle 8(\sqrt{2}/2), 8(\sqrt{2}/2) \rangle$
$= \langle 4\sqrt{2}, 4\sqrt{2} \rangle$

41. $\langle 290\cos 145°, 290\sin 145° \rangle = \langle -237.6, 166.3 \rangle$

43. $\langle 18\cos 347°, 18\sin 347° \rangle = \langle 17.5, -4.0 \rangle$

45. $\langle 15, -10 \rangle$

47. $\langle 6, -4 \rangle + \langle 12, -18 \rangle = \langle 18, -22 \rangle$

49. $\langle -1, 5 \rangle + \langle 12, -18 \rangle = \langle 11, -13 \rangle$

51. $\langle 3, -2 \rangle - \langle 3, -1 \rangle = \langle 0, -1 \rangle$

53. $(3)(-1) + (-2)(5) = -13$

55. If $A = \langle 2, 1 \rangle$ and $B = \langle 3, 5 \rangle$, then the angle between these vectors is given by

$$\cos^{-1}\left(\frac{A \cdot B}{|\,A\,| \cdot |\,B\,|}\right) = \cos^{-1}\left(\frac{11}{\sqrt{5}\sqrt{34}}\right) \approx 32.5°$$

57. If $A = \langle -1, 5 \rangle$ and $B = \langle 2, 7 \rangle$, then the angle between these vectors is given by

$$\cos^{-1}\left(\frac{A \cdot B}{|A| \cdot |B|}\right) = \cos^{-1}\left(\frac{33}{\sqrt{26}\sqrt{53}}\right) \approx 27.3°$$

59. Since $\langle -6, 5 \rangle \cdot \langle 5, 6 \rangle = 0$, the angle between them is $90°$.

61. Perpendicular since their dot product is zero

63. Parallel since $-2\langle 1, 7 \rangle = \langle -2, -14 \rangle$

65. Neither

67. $2i + j$ **69.** $-3i + \sqrt{2}j$ **71.** $-9j$

73. $-7i - j$

75. The magnitude of $A + B = \langle 1, 4 \rangle$ is

$$\sqrt{1^2 + 4^2} = \sqrt{17}$$

and the direction angle is

$$\tan^{-1}(4/1) \approx 76.0°$$

77. The magnitude of $-3A = \langle -9, -3 \rangle$ is

$$\sqrt{(-9)^2 + (-3)^2} = \sqrt{90} = 3\sqrt{10}.$$

Since $\tan^{-1}(3/9) \approx 18.4°$, the direction angle is
$$180° + 18.4° = 198.4°.$$

79. The magnitude of $B - A = \langle -5, 2 \rangle$ is

$$\sqrt{(-5)^2 + 2^2} = \sqrt{29}.$$

Since $\tan^{-1}(-2/5) \approx -21.8°$, the direction angle is
$$180° - 21.8° = 158.2°.$$

81. Note $-A + \frac{1}{2}B = \langle -3 - 1, -1 + 3/2 \rangle$
$= \langle -4, 1/2 \rangle$. The magnitude is
$$\sqrt{(-4)^2 + (1/2)^2} = \sqrt{65}/2.$$
Since $\tan^{-1}\left(\frac{1/2}{-4}\right) \approx -7.1°$, the direction angle is $180° - 7.1° = 172.9°$.

83. The resultant is $\langle 2 + 6, 3 + 2 \rangle = \langle 8, 5 \rangle$. Then the magnitude is

$$\sqrt{8^2 + 5^2} = \sqrt{89}$$

and direction angle is

$$\tan^{-1}(5/8) = 32.0°.$$

85. The resultant is $\langle -6 + 4, 4 + 2 \rangle = \langle -2, 6 \rangle$ and its magnitude is

$$\sqrt{(-2)^2 + 6^2} = 2\sqrt{10}.$$

Since $\tan^{-1}(-6/2) \approx -71.6°$, the direction angle is
$$180° - 71.6° = 108.4°.$$

87. The resultant is $\langle -4 + 3, 4 - 6 \rangle = \langle -1, -2 \rangle$ and its magnitude is $\sqrt{(-1)^2 + (-2)^2} = \sqrt{5}$. Since $\tan^{-1}(2/1) \approx 63.4°$, the direction angle is $180° + 63.4° = 243.4°$.

89. The magnitudes of the horizontal and vertical components are $|520\cos 30°| \approx 450.3$ mph and $|520\sin 30°| = 260$ mph, respectively.

93. Apply Heron's formula with $a = b = 3$ and $c = 1$. Let $s = (a + b + c)/2 = 7/2$. The area is

$$\sqrt{s(s-a)(s-b)(s-c)} = \sqrt{\frac{7}{2} \cdot \frac{1}{2} \cdot \frac{1}{2} \cdot \frac{5}{2}}$$
$$= \sqrt{\frac{35}{16}}$$
$$= \frac{\sqrt{35}}{4}.$$

95. a) $\dfrac{\tan x + \tan y}{1 - \tan x \tan y}$

b) $\dfrac{\tan x - \tan y}{1 + \tan x \tan y}$

97. Other acute angle is $90° - 33° = 57°$, other legs are $66\sin 33° \approx 35.9$ ft, and $66\cos 33° \approx 55.4$ ft

99. **a)** Consider the top layer of four balls and the ball that sits above it. Connecting the centers of the five balls gives a pyramid with a 2-by-2 square base and a slanted height of 2. The diagonal of the square base is $2\sqrt{2}$. By looking at a slanted height and half of the base, the height of the pyramid is $\sqrt{2}$. Then the height of the box is $4 + \sqrt{2}$. Multiplying by the area of the 4-by-4 base of the box which is 16, we obtain the volume of the box. That is, the volume of the box is

$$16(4 + \sqrt{2}) = 64 + 16\sqrt{2} \approx 86.63.$$

b) Using the same pyramid in part a), we obtain that the height of the box is $2 + 2\sqrt{2}$. Multiplying by the area of the 4-by-4 base of the box which is 16, we find that the volume of the box is

$$16(2 + 2\sqrt{2}) = 32 + 32\sqrt{2} \approx 77.25.$$

c) The distance between a vertex of the cubic box and the center of the ball that is closest to the vertex is $\sqrt{3}$. This is obtained by using the Pythagorean theorem.

Then the diagonal of the cube holding the five balls has a length of

$$4 + 2\sqrt{3}.$$

If x is the length of an edge of the cube then

$$x^2 + x^2 + x^2 = (4 + 2\sqrt{3})^2.$$

Solving for x, we find $x = \dfrac{4 + 2\sqrt{3}}{\sqrt{3}}$. Then the volume of the cube is

$$\text{Volume} = x^3 = \left(\frac{4 + 2\sqrt{3}}{\sqrt{3}}\right)^3.$$

Simplifying, we obtain

$$\text{Volume} = \frac{208\sqrt{3}}{9} + 40 \approx 80.03.$$

For Thought

1. True, since the force required is

$$99\sin 88° \approx 98.9 \text{ kg.}$$

2. True

3. False, the weight of an object is modelled by a vertical vector.

4. True **5.** True **6.** True

7. False, the bearing of the wind is 45°.

8. False, the airplane's ground speed is slower than 400 mph since the airplane is flying against the wind.

9. False, the bearing of the course is 185°.

10. True, for the bearing of the plane's course is

$$135° + 3° = 138°.$$

5.5 Exercises

1. Draw two perpendicular vectors whose magnitudes are 3 and 8.

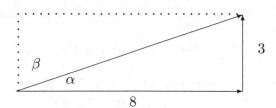

The magnitude of the resultant force is

$$\sqrt{8^2 + 3^2} = \sqrt{73} \approx 8.5 \text{ lb}$$

by the Pythagorean Theorem.

The angles between the resultant and each force are

$$\tan^{-1}(3/8) \approx 20.6°$$

and

$$\beta = 90° - 20.6° = 69.4°.$$

3. Draw two vectors with magnitudes 10.3 and 4.2 that act at an angle of 130° with each other.

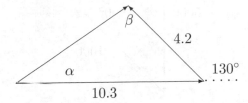

By using the cosine law, the magnitude of the resultant force is

$$r = \sqrt{10.3^2 + 4.2^2 - 2(10.3)(4.2)\cos 50°}$$
$$\approx 8.3 \text{ newtons.}$$

By the sine law, we find

$$\frac{4.2}{\sin \alpha} = \frac{r}{\sin 50°}$$
$$\sin \alpha = \frac{4.2 \sin 50°}{r}$$
$$\sin \alpha \approx 0.3898$$
$$\alpha \approx \sin^{-1}(0.3898) \approx 22.9°.$$

The angles between the resultant and each force are 22.9° and

$$\beta = 180° - 22.9° - 50° = 107.1°.$$

5. Draw two vectors with magnitudes 10 & 12.3 and whose angle between them is 23.4°.

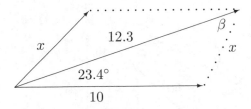

By the cosine law, the magnitude of the other force is
$$x = \sqrt{10^2 + 12.3^2 - 2(10)(12.3)\cos 23.4°}$$
$$\approx 5.051 \approx 5.1 \text{ pounds.}$$

By the sine law, we obtain

$$\frac{10}{\sin \beta} = \frac{5.051}{\sin 23.4°}$$
$$\sin \beta = \frac{10 \sin 23.4°}{5.051}$$
$$\sin \beta \approx 0.7863$$
$$\beta \approx \sin^{-1}(0.7863) \approx 51.8°.$$

The angle between the two forces is

$$51.8° + 23.4° = 75.2°.$$

7. Since the angles in a parallelogram must add up to 360°, the angle formed by the two forces is

$$\frac{360° - 2(25°)}{2} = 155°.$$

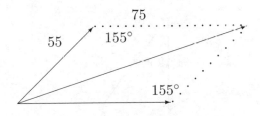

By the cosine law, the magnitude of the resultant force is

$$\sqrt{55^2 + 75^2 - 2(55)(75)\cos 155°} \approx 127.0 \text{ pounds.}$$

Then the donkey must pull a force of 127 pounds in the direction opposite that of the resultant's direction.

9. If x is the amount of force required as shown on the next page, then

$$\frac{x}{3000} = \sin 20°$$
$$x = 3000 \sin 20°$$
$$x \approx 1026.1 \text{ lb}$$

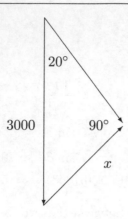

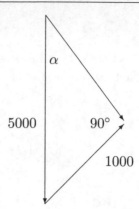

11. If w is the weight of the block of ice as shown below, then

$$\sin 25° = \frac{100}{w}$$

$$w = \frac{100}{\sin 25°}$$

$$w \approx 236.6 \text{ lb}$$

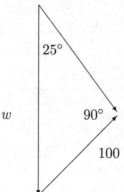

13. If α is the angle of inclination of the hill as shown on the next column, then

$$\sin \alpha = \frac{1000}{5000}$$

$$\alpha = \sin^{-1} \frac{1}{5}$$

$$\alpha \approx 11.5°$$

15. Let x be the ground speed and let α be drift angle as shown below.

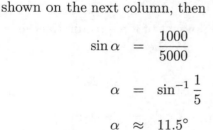

By the Pythagorean Theorem, we obtain

$$x = \sqrt{240^2 + 30^2} \approx 241.9 \text{ mph}$$

Using right triangle trigonometry, we obtain

$$\alpha = \tan^{-1} \frac{30}{240} \approx 7.1°.$$

Thus, the bearing of the course is

$$90° + \alpha \approx 97.1°.$$

17. Let x be the ground speed and let α be drift angle as shown below.

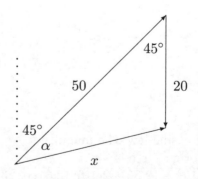

Applying the cosine law, we obtain

$$x = \sqrt{20^2 + 50^2 - 2(20)(50)\cos 45°}$$
$$\approx 38.5 \text{ mph}.$$

Using the sine law, we find

$$\frac{\sin \alpha}{20} = \frac{\sin 45°}{x}$$

$$\alpha = \sin^{-1}\left(\frac{20 \sin 45°}{x}\right)$$

$$\alpha \approx 21.5°.$$

Thus, the bearing of the course is

$$45° + \alpha \approx 66.5°.$$

19. Let x be the ground speed and let α be drift angle as shown below.

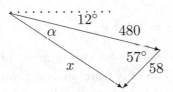

Note, the angle between the vectors representing the airplane and the wind is

$$57° = 12° + 45°.$$

By the cosine law, the ground speed is

$$x = \sqrt{480^2 + 58^2 - 2(480)(58)\cos 57°}$$
$$\approx 451.0 \text{ mph}.$$

By the sine law, we get

$$\frac{\sin \alpha}{58} = \frac{\sin 57°}{x}$$

$$\sin \alpha = \frac{58 \sin 57°}{x}$$

$$\alpha = \sin^{-1}\left(\frac{58 \sin 57°}{x}\right)$$

$$\alpha \approx 6.2°.$$

The bearing of the airplane is

$$102° + \alpha \approx 108.2°.$$

21. Draw two vectors representing the canoe and river current; the magnitudes of these vectors are 2 and 6, respectively.

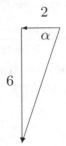

Since $\alpha = \tan^{-1}(6/2) \approx 71.6°$, the direction measured from the north is

$$270° - 71.6° = 198.4°.$$

Also, if d is the distance downstream from a point directly across the river to the point where she will land, then $\tan \alpha = d/2000$. Since $\tan \alpha = 6/2 = 3$, we get

$$d = 2000 \cdot 3 = 6000 \text{ ft}.$$

23. a) Assume we have a coordinate system where the origin is the point where the boat will start.

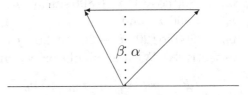

The intended direction and speed of the boat that goes 3 mph in still water is defined by the vector $3 \sin \alpha \, \boldsymbol{i} + 3 \cos \alpha \, \boldsymbol{j}$ and its actual direction and speed is determined by the vector

$$\boldsymbol{v} = (3 \sin \alpha - 1) \, \boldsymbol{i} + 3 \cos \alpha \, \boldsymbol{j}.$$

The number t of hours it takes the boat to cross the river is given by

$$t = \frac{0.2}{3 \cos \alpha},$$

the solution to $3t \cos \alpha = 0.2$.
Suppose $\beta > 0$ if $3 \sin \alpha - 1 < 0$ and
$\beta < 0$ if $3 \sin \alpha - 1 > 0$. Using right
triangle trigonometry, we find

$$\tan \beta = \frac{|3 \sin \alpha - 1|}{3 \cos \alpha}.$$

The distance d the boat travels as a
function of β is given by

$$\begin{aligned}
d &= |t\, \boldsymbol{v}| \\
&= \frac{0.2}{3 \cos \alpha} |\boldsymbol{v}| \\
&= \frac{0.2}{3 \cos \alpha} \sqrt{(3 \sin \alpha - 1)^2 + (3 \cos \alpha)^2} \\
&= 0.2\sqrt{\tan^2 \beta + 1} \\
d &= 0.2|\sec \beta|.
\end{aligned}$$

b) Since speed is distance divided by time,
then by using the answer from part a)
the speed r as a function of α and β is

$$\begin{aligned}
r &= \frac{d}{t} \\
&= \frac{0.2|\sec \beta|}{0.2/(3 \cos \alpha)} \\
r &= 3 \cos(\alpha)|\sec \beta|.
\end{aligned}$$

25. Let the forces exerted by the papa, mama, and
baby elephant be represented by the vectors
$\boldsymbol{v_p} = 800 \cos 30° \, \boldsymbol{i} + 800 \sin 30° \, \boldsymbol{j}$,
$\boldsymbol{v_m} = 500 \, \boldsymbol{i}$, and
$\boldsymbol{v_b} = 200 \cos 20° \, \boldsymbol{i} - 200 \sin 20° \, \boldsymbol{j}$,
respectively. With a calculator, we find

$$\begin{aligned}
\boldsymbol{F} &= \boldsymbol{v_p} + \boldsymbol{v_m} + \boldsymbol{v_b} \\
&\approx 1380.76 \, \boldsymbol{i} + 331.60 \, \boldsymbol{j}.
\end{aligned}$$

The magnitude of the resultant of the three
forces is

$$|\boldsymbol{F}| \approx \sqrt{1380.76^2 + 331.60^2} \approx 1420.0 \text{ lb}.$$

27. By the cosine law,

$$\begin{aligned}
\gamma &= \cos^{-1}\left(\frac{a^2 + b^2 - c^2}{2ab}\right) \\
&= \cos^{-1}\left(\frac{-26}{70}\right) \\
&\approx 111.8°
\end{aligned}$$

By the sine law,

$$\beta = \sin^{-1}\left(\frac{7 \sin \gamma}{10}\right) \approx 40.5°$$

Then $\alpha = 180° - \gamma - \beta \approx 27.7°$.

29. Let $\boldsymbol{a} = \langle -3, 5 \rangle$ and $\boldsymbol{b} = \langle 1, 6 \rangle$. If α is the
smallest positive angle between the two vec-
tors, then

$$\begin{aligned}
\boldsymbol{a} \cdot \boldsymbol{b} &= |\boldsymbol{a}| \cdot \boldsymbol{b} \cdot \cos \alpha \\
27 &= \sqrt{34} \cdot \sqrt{37} \cdot \cos \alpha \\
\alpha &= \cos^{-1}\left(\frac{27}{\sqrt{34} \cdot \sqrt{37}}\right) \\
\alpha &\approx 40.4°
\end{aligned}$$

31. $230 \tan 48° \approx 255.4$ ft

33. Let $1 \leq x \leq 199$ be the number of dogs remaining. The only number less than 200 that
is divisible by four of the five denominators
below

$$3, 4, 5, 7, \text{ and } 9$$

is $x = 180$. Then there are $180/9 = 20$
dachshunds remaining, $3(20) = 60$ original
beagles, $180/5 = 36$ beagles remaining, and
$60 - 36 = 24$ beagles escaped.

Chapter 5 Review Exercises

1. Draw a triangle with $\gamma = 48°$, $a = 3.4$, $b = 2.6$.

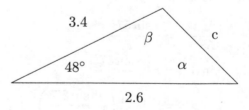

By the cosine law, we obtain
$c = \sqrt{2.6^2 + 3.4^2 - 2(2.6)(3.4) \cos 48°} \approx$
$2.5475 \approx 2.5$. By the sine law, we find

$$\frac{2.5475}{\sin 48°} = \frac{2.6}{\sin \beta}$$

$$\sin \beta = \frac{2.6 \sin 48°}{2.5475}$$

$$\sin \beta \approx 0.75846$$
$$\beta \approx \sin^{-1}(0.75846)$$
$$\beta \approx 49.3°.$$

Also, $\alpha = 180° - (49.3° + 48°) = 82.7°$.

3. Draw a triangle with $\alpha = 13°$, $\beta = 64°$, $c = 20$.

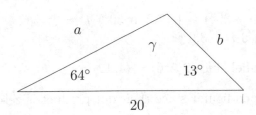

Note $\gamma = 180° - (64° + 13°) = 103°$.

By the sine law, we get $\dfrac{20}{\sin 103°} = \dfrac{a}{\sin 13°}$

and $\dfrac{20}{\sin 103°} = \dfrac{b}{\sin 64°}$.

So $a = \dfrac{20}{\sin 103°} \sin 13° \approx 4.6$

and $b = \dfrac{20}{\sin 103°} \sin 64° \approx 18.4$.

5. Draw a triangle with $a = 3.6$, $b = 10.2$, $c = 5.9$.

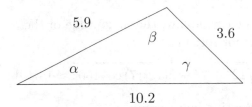

By the cosine law one gets

$$\cos \beta = \frac{5.9^2 + 3.6^2 - 10.2^2}{2(5.9)(3.6)} \approx -1.3.$$

This is a contradiction since the range of cosine is $[-1, 1]$. No triangle exists.

7. Draw a triangle with sides $a = 30.6$, $b = 12.9$, and $c = 24.1$.

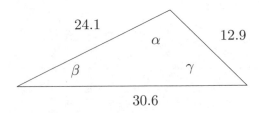

By the cosine law, we get

$$\cos \alpha = \frac{24.1^2 + 12.9^2 - 30.6^2}{2(24.1)(12.9)} \approx -0.3042.$$

So $\alpha = \cos^{-1}(-0.3042) \approx 107.7°$.

Similarly, we find

$$\cos \beta = \frac{24.1^2 + 30.6^2 - 12.9^2}{2(24.1)(30.6)} \approx 0.9158.$$

So $\beta = \cos^{-1}(0.9158) \approx 23.7°$.
Also, $\gamma = 180° - (107.7° + 23.7°) = 48.6°$.

9. Draw angle $\beta = 22°$ and let h be the height.

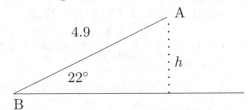

Since $h = 4.9 \sin 22° \approx 1.8$ and $1.8 < b < 4.9$, we have two triangles as follows:

Case 1:

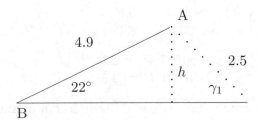

and

Case 2:

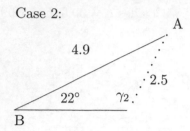

Apply the sine law to case 1.

$$\frac{4.9}{\sin \gamma_1} = \frac{2.5}{\sin 22°}$$

$$\sin \gamma_1 = \frac{4.9 \sin 22°}{2.5}$$

$$\sin \gamma_1 \approx 0.7342$$

$$\gamma_1 = \sin^{-1}(0.7342) \approx 47.2°$$

So $\alpha_1 = 180° - (22° + 47.2°) = 110.8°$.

By the sine law, $a_1 = \dfrac{2.5}{\sin 22°} \sin 110.8° \approx 6.2$.

In case 2, $\gamma_2 = 180° - \gamma_1 = 132.8°$
and $\alpha_2 = 180° - (22° + 132.8°) = 25.2°$.

By the sine law, $a_2 = \dfrac{2.5}{\sin 22°} \sin 25.2° \approx 2.8$.

11. Area is $A = \dfrac{1}{2}(12.2)(24.6)\sin 38° \approx 92.4$ ft^2.

13. Since $S = \dfrac{5.4 + 12.3 + 9.2}{2} = 13.45$, the
area is

$$\sqrt{13.45(13.45 - 5.4)(13.45 - 12.3)(13.45 - 9.2)}$$
$$\approx 23.0 \text{ km}^2.$$

15. $|\, v_x \,| = |6 \cos 23.3°| \approx 5.5$,
$|\, v_y \,| = |6 \sin 23.3°| \approx 2.4$

17. $|\, v_x \,| = |3.2 \cos 231.4°| \approx 2.0$,
$|\, v_y \,| = |3.2 \sin 231.4°| \approx 2.5$

19. magnitude $\sqrt{2^2 + 3^2} = \sqrt{13}$, direction angle
$\tan^{-1}(3/2) \approx 56.3°$

21. The magnitude is $\sqrt{(-3.2)^2 + (-5.1)^2}$
≈ 6.0. Since $\tan^{-1}(5.1/3.2) \approx 57.9°$, the
direction angle is $180° + 57.9° = 237.9°$.

23. $\langle \sqrt{2} \cos 45°, \sqrt{2} \sin 45° \rangle = \langle 1, 1 \rangle$

25. $\langle 9.1 \cos 109.3°, 9.1 \sin 109.3° \rangle \approx \langle -3.0, 8.6 \rangle$

27. $\langle -6, 8 \rangle$

29. $\langle 2 - 2, -5 - 12 \rangle = \langle 0, -17 \rangle$

31. $\langle -1, 5 \rangle \cdot \langle 4, 2 \rangle = -4 + 10 = 6$

33. $-4\, \boldsymbol{i} + 8\, \boldsymbol{j}$

35. $(7.2 \cos 30°)\, \boldsymbol{i} + (7.2 \sin 30°)\, \boldsymbol{j} \approx$
$3.6\sqrt{3}\, \boldsymbol{i} + 3.6\, \boldsymbol{j}$

37. Parallel since $2\langle 2, 6 \rangle = \langle 4, 12 \rangle$

39. Perpendicular since their dot product is zero

41. Parallel since $-3\langle -3, 8 \rangle = \langle 9, -24 \rangle$

43. Draw two vectors with magnitudes 7 and 12
that act at an angle of 30° with
each other.

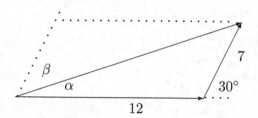

By the cosine law, the magnitude of the
resultant force is

$$\sqrt{12^2 + 7^2 - 2(12)(7) \cos 150°} \approx 18.4 \text{ lb}.$$

By the sine law, we find

$$\frac{7}{\sin \alpha} = \frac{18.4}{\sin 150°}$$

$$\sin \alpha = \frac{7 \sin 150°}{18.4}$$

$$\sin \alpha \approx 0.19$$

$$\alpha \approx \sin^{-1}(0.19) \approx 11.0°.$$

The angles between the resultant and
the two forces are 11.0° and
$\beta = 180° - 150° - 11° = 19.0°$.

45. Let x be the ground speed and let α be the drift angle, as shown below.

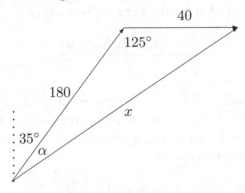

Note, the angle between the vectors of the plane and the wind is

$$125° = 90° + 35°.$$

Applying the cosine law, we obtain

$$x = \sqrt{180^2 + 40^2 - 2(180)(40)\cos 125°}$$

$$x \approx 205.6 \text{ mph}.$$

Applying the sine law, we find

$$\frac{\sin\alpha}{40} = \frac{\sin 125°}{x}$$

$$\alpha = \sin^{-1}\left(\frac{40\sin 125°}{x}\right)$$

$$\alpha \approx 9.2°.$$

The bearing of the plane's course is

$$35° + \alpha \approx 44.2°.$$

47. Using Heron's formula and since

$$\frac{482 + 364 + 241}{2} = 543.5,$$

the area of Susan's lot is

$$\sqrt{543.5(543.5 - 482)(543.5 - 364)(543.5 - 241)}$$

which is approximately

$$42,602 \text{ ft}^2.$$

Similarly, since

$$\frac{482 + 369 + 238}{2} = 544.5,$$

the area of Seth's lot is

$$\sqrt{544.5(544.5 - 482)(544.5 - 369)(544.5 - 238)}$$

or approximately

$$42,785 \text{ ft}^2.$$

Then Seth got the larger piece.

49. Consider triangle below.

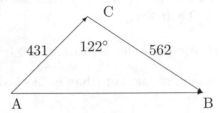

The distance between A and B is

$$\sqrt{431^2 + 562^2 - 2(431)(562)\cos 122°} \approx 870.82 \text{ ft}.$$

The extra amount spent is

$$(431 + 562 - 870.82)(\$21.60) \approx \$2639.$$

51. Let α be the base angle of the larger isosceles triangle. Drop a perpendicular from the top vertex to the base.

The perpendicular bisects the base of unit length into two equal parts.

Using right triangle trigonometry, we find $\cos\alpha = \frac{1}{4}$. The area of the shaded triangle is

$$\begin{aligned}
\text{Area} &= \frac{1}{2}ab\sin C \\
&= \frac{1}{2}\sin\alpha \\
&= \frac{1}{2}\sqrt{1 - \frac{1}{16}} \\
\text{Area} &= \frac{\sqrt{15}}{8}
\end{aligned}$$

53. Let r be the radius of the sun and earth, which we assume are equal. Let a and α be the length of a chord and the corresponding central angle, respectively.

We are given $a = 0.8(2r)$. Then the length of the chord satisfies

$$
\begin{aligned}
a &= r\sqrt{2 - 2\cos\alpha} \\
1.6r &= r\sqrt{2 - 2\cos\alpha} \\
1.6^2 &= 2 - 2\cos\alpha \\
\cos\alpha &= -0.28.
\end{aligned}
$$

The part of the sun that is blocked is two times the area of a lens-like region. See Exercise 33c, Section 5.3. Then

$$(\text{Twice Area of Lens}) = r^2(\alpha - \sin\alpha)$$

The percentage of the sun that is blocked is given by

$$\frac{\text{Twice Area of Lens}}{\pi r^2} = \frac{\alpha - \sin\alpha}{\pi} \approx 28\%.$$

55. Put the moon's center initially at $(0,0)$. Assume the moon and sun are circles of radii 1. Put the sun's center at $(2,0)$. At time $t = 1$, put the moon's center at $(4,0)$. Then at time t, the moons center is at $C_t = (4t, 0)$.

At time t, the moon and sun intersects at some point P_t in the first quadrant. Drop a perpendicular from P_t to the x-axis. Label the foot of this perpendicular by F_t which is a point on the x-axis.

Let $\alpha_t = \angle P_t C_t F_t$ be the angle at C_T in the right triangle $\triangle P_t C_t F_t$. Notice, $\overline{C_t F_t} = 1 - 2t$.

Using right triangle trigonometry,

$$\cos\frac{\alpha_t}{2} = 1 - 2t.$$

By a double-angle identity for cosine,

$$
\begin{aligned}
\cos\alpha_t &= 2\cos^2\alpha_t - 1 \\
&= 2(1 - 2t)^2 - 1 \\
&= 1 - 8t + 8t^2.
\end{aligned}
$$

The area of the sun that is blocked is two times the area of a lens-like region.

$$
\begin{aligned}
(\text{Twice Area of Lens}) &= r^2(\alpha_t - \sin\alpha_t) \\
&= \cos^{-1}u - \sin(\cos^{-1}u).
\end{aligned}
$$

where $r = 1$, and $u = 1 - 8t + 8t^2$. Hence, the portion of the sun that is blocked is the above area divided by the area of the sun, i.e.,

$$\frac{\cos^{-1}u - \sin(\cos^{-1}u)}{\pi}.$$

57. The sides of the three squares are $\sqrt{8}$, $\sqrt{13}$, and $\sqrt{17}$. These are also the sides of the triangle. We use Heron's formula to find the area of the triangle. Let

$$s = \frac{\sqrt{8} + \sqrt{13} + \sqrt{17}}{2}.$$

The area of the triangle is

$$
\begin{aligned}
\text{Area} &= \sqrt{s(s - \sqrt{8})(s - \sqrt{13})(s - \sqrt{17})} \\
&= 5 \text{ acres} \\
&= 5(43,560) \text{ ft}^2 \\
\text{Area} &= 217,800 \text{ ft}^2.
\end{aligned}
$$

Chapter 5 Test

1. Draw a triangle with $\alpha = 30°$, $b = 4$, $a = 2$.

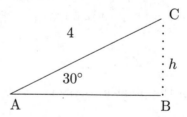

Since $h = 4\sin 30° = 2$ and $a = 2$, there is only one triangle and $\beta = 90°$. Then $\gamma = 90° - 30° = 60°$. Since $c^2 + 2^2 = 4^2$, we get $c = \sqrt{12} = 2\sqrt{3}$.

2. Draw angle $\alpha = 60°$ and let h be the height.

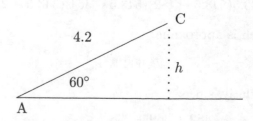

Since $h = 4.2 \sin 60° \approx 3.6$ and $3.6 < a < 4.2$, there are two triangles and they are given by

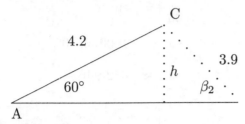

and

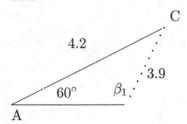

Apply the sine law to the acute triangle.

$$\frac{3.9}{\sin 60°} = \frac{4.2}{\sin \beta_2}$$

$$\sin \beta_2 = \frac{4.2 \sin 60°}{3.9}$$

$$\sin \beta_2 \approx 0.93264$$

$$\beta_2 = \sin^{-1}(0.93264) \approx 68.9°$$

So $\gamma_2 = 180° - (\beta_2 + 60°) = 51.1°$.

By the sine law, $c_2 = \dfrac{3.9}{\sin 60°} \sin 51.1° \approx 3.5$.

In the obtuse triangle, $\beta_1 = 180° - \beta_2 = 111.1°$ and $\gamma_1 = 180° - (\beta_1 + 60°) = 8.9°$.

By the sine law, $c_1 = \dfrac{3.9}{\sin 60°} \sin 8.9° \approx 0.7$.

3. Draw the only triangle with $a = 3.6$, $\alpha = 20.3°$, and $\beta = 14.1°$.

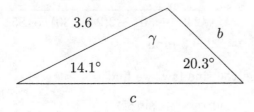

Note, $\gamma = 180° - 14.1° - 20.3° = 145.6°$. Using the sine law, we find

$$\frac{b}{\sin 14.1°} = \frac{3.6}{\sin 20.3°} \text{ and } \frac{c}{\sin 145.6°} = \frac{3.6}{\sin 20.3°}.$$

Then $b = \dfrac{3.6}{\sin 20.3°} \sin 14.1° \approx 2.5$ and

$$c = \frac{3.6}{\sin 20.3°} \sin 145.6° \approx 5.9.$$

4. Draw the only triangle with $a = 2.8$, $b = 3.9$, and $\gamma = 17°$.

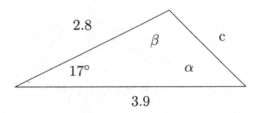

By the cosine law, we get
$c = \sqrt{3.9^2 + 2.8^2 - 2(3.9)(2.8)\cos 17°} \approx 1.47 \approx 1.5$. By the sine law,

$$\frac{1.47}{\sin 17°} = \frac{2.8}{\sin \alpha}$$

$$\sin \alpha = \frac{2.8 \sin 17°}{1.47}$$

$$\sin \alpha \approx 0.5569$$

$$\alpha \approx \sin^{-1}(0.5569) \approx 33.8°.$$

Also, $\beta = 180° - (33.8° + 17°) = 129.2°$.

5. Draw the only triangle with the given sides $a = 4.1$, $b = 8.6$, and $c = 7.3$.

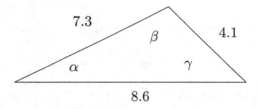

First, find the largest angle β by the cosine law.

$$\cos \beta = \frac{7.3^2 + 4.1^2 - 8.6^2}{2(7.3)(4.1)}$$

$$\cos \beta \approx -0.06448$$

$$\beta \approx \cos^{-1}(-0.06448)$$

$$\beta \approx 93.7°.$$

By the sine law,

$$\frac{8.6}{\sin 93.7°} = \frac{7.3}{\sin \gamma}$$

$$\sin \gamma = \frac{7.3 \sin 93.7°}{8.6}$$

$$\sin \gamma \approx 0.8471$$

$$\gamma \approx \sin^{-1}(0.8471) \approx 57.9°.$$

Also, $\alpha = 180° - (57.9° + 93.7°) = 28.4°.$

6. The magnitude of $\mathbf{A} + \mathbf{B} = \langle -2, 6 \rangle$ is

$$\sqrt{(-2)^2 + 6^2} = \sqrt{40} = 2\sqrt{10}.$$

The direction angle is

$$\cos^{-1}(-2/\sqrt{40}) \approx 108.4°.$$

7. The magnitude of $\mathbf{A} - \mathbf{B} = \langle -4, -2 \rangle$ is

$$\sqrt{(-4)^2 + (-2)^2} = \sqrt{20} = 2\sqrt{5}.$$

Since $\tan^{-1}(2/4) \approx 26.6°$, the direction angle is

$$180° + 26.6° = 206.6°.$$

8. The magnitude of $3\mathbf{B} = \langle 3, 12 \rangle$ is

$$\sqrt{3^2 + 12^2} = \sqrt{153} = 3\sqrt{17}.$$

The direction angle is

$$\tan^{-1}(12/3) \approx 76.0°.$$

9. The area is $\dfrac{1}{2}(12)(10)\sin(22°) \approx 22.5$ ft^2.

10. Using Heron's formula and if

$$s = \frac{4.1 + 6.8 + 9.5}{2} = 10.2$$

then the area is

$$\sqrt{s(s - 4.1)(s - 6.8)(s - 9.5)} \approx 12.2 \text{ m}^2.$$

11. Since $a_1 = 4.6 \cos 37.2° \approx 3.66$ and $a_2 = 4.6 \sin 37.2° \approx 2.78$, we have

$$\mathbf{v} \approx 3.66 \, \mathbf{i} + 2.78 \, \mathbf{j} \, .$$

12. Perpendicular since their dot product is zero. That is, $\langle -3, 5 \rangle \cdot \langle 5, 3 \rangle = (-3)(5) + (5)(3) = 0.$

13. If x is the force required to push the riding lawnmower as shown below, then

$$\frac{x}{1000} = \sin 40°$$

$$x = 1000 \sin 40°$$

$$x \approx 642.8 \text{ lb}$$

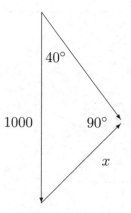

14. Let x be the ground speed and let α be drift angle as shown below.

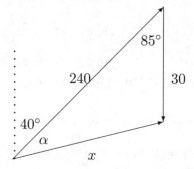

Note, the angle between the vector of the airplane and the vector of the wind is

$$85° = 40° + 45°.$$

Applying the cosine law, we obtain

$$x = \sqrt{240^2 + 30^2 - 2(240)(30)\cos 85°}$$
$$\approx 239.3 \text{ mph}.$$

Using the sine law, we find

$$\frac{\sin \alpha}{30} = \frac{\sin 85°}{x}$$

$$\alpha = \sin^{-1}\left(\frac{30\sin 85°}{x}\right)$$

$$\alpha \approx 7.2°.$$

Thus, the bearing of the course is

$$40° + \alpha \approx 47.2°.$$

Tying It All Together

1. $\sin(\pi/6) = 1/2$, $\cos(\pi/6) = \sqrt{3}/2$, $\tan(\pi/6) = \sqrt{3}/3$, $\csc(\pi/6) = 2$, $\sec(\pi/6) = 2\sqrt{3}/3$, and $\cot(\pi/6) = \sqrt{3}$

2. $\sin(\pi/4) = \sqrt{2}/2$, $\cos(\pi/4) = \sqrt{2}/2$, $\tan(\pi/4) = 1$, $\csc(\pi/4) = \sqrt{2}$, $\sec(\pi/4) = \sqrt{2}$, and $\cot(\pi/4) = 1$

3. $\sin(\pi/3) = \sqrt{3}/2$, $\cos(\pi/3) = 1/2$, $\tan(\pi/3) = \sqrt{3}$, $\csc(\pi/3) = 2\sqrt{3}/3$, $\sec(\pi/3) = 2$, and $\cot(\pi/3) = \sqrt{3}/3$

4. $\sin(\pi/2) = 1$, $\cos(\pi/2) = 0$, $\tan(\pi/2)$ is undefined, $\csc(\pi/2) = 1$, $\sec(\pi/2)$ is undefined, and $\cot(\pi/2) = 0$

5. $\pi/2$ 6. $-\pi/2$ 7. $-\pi/6$ 8. $\pi/6$

9. π 10. 0 11. $5\pi/6$ 12. $\pi/6$

13. 0 14. $\pi/4$ 15. $-\pi/4$ 16. $\pi/6$

17. $\{x \mid x = k\pi \text{ where } k \text{ is an integer}\}$

18. Factoring, we get

$$\sin(x)(\sin(x) - 1) = 0.$$

Then
$$\sin(x) = 0 \text{ or } \sin(x) = 1.$$
Thus, the solution set is

$$\{x \mid x = k\pi \text{ or } x = \frac{\pi}{2} + 2k\pi\}.$$

19. Factoring, we obtain

$$\sin^2 x - \sin x - 2 = 0$$
$$(\sin x + 1)(\sin x - 2) = 0.$$

Then
$$\sin x = -1 \text{ or } \sin x = 2.$$

Since $\sin x = 2$ is impossible, we have

$$\sin x = -1.$$

The solution set is

$$\left\{x \mid x = \frac{3\pi}{2} + 2k\pi\right\}.$$

20. Factoring, we find

$$4\sin x \cos x - 2\cos x + 2\sin x - 1 = 0$$
$$2\cos x(2\sin x - 1) + (2\sin x - 1) = 0$$
$$(2\cos x + 1)(2\sin x - 1) = 0.$$

Then
$$\cos x = -\frac{1}{2} \text{ or } \sin x = \frac{1}{2}.$$

The solution set is $\left\{x \mid x = \dfrac{\pi}{6} + 2k\pi,\right.$

$\left. x = \dfrac{5\pi}{6} + 2k\pi, x = \dfrac{2\pi}{3} + 2k\pi, x = \dfrac{4\pi}{3} + 2k\pi\right\}.$

21. Factoring, we find

$$4x\sin x + 2\sin x - 2x - 1 = 0$$
$$2\sin x(2x + 1) - (2x + 1) = 0$$
$$(2x + 1)(2\sin x - 1) = 0.$$

Then
$$x = -\frac{1}{2} \text{ or } \sin x = \frac{1}{2}.$$

The solution set is
$$\left\{x \mid x = -\frac{1}{2}, x = \frac{\pi}{6} + 2k\pi, x = \frac{5\pi}{6} + 2k\pi\right\}.$$

22. Since $\sin 2x = 1/2$, we obtain

$$2x = \frac{\pi}{6} + 2k\pi \text{ or } 2x = \frac{5\pi}{6} + 2k\pi$$

where k is an integer. Then the solution set is

$$\left\{x \mid x = \frac{\pi}{12} + k\pi \text{ or } x = \frac{5\pi}{12} + k\pi\right\}.$$

23. Since $\tan 4x = 1/\sqrt{3}$, we obtain

$$4x = \frac{\pi}{6} + k\pi$$

where k is an integer. Then the solution set is

$$\left\{x \mid x = \frac{\pi}{24} + \frac{k\pi}{4}\right\}.$$

24. Since

$$\sin^2 x + \cos^2 x = 1$$

is an identity, the solution set is the set of all real numbers.

25. Amplitude 1, period $2\pi/3$, phase shift 0, domain $(-\infty, \infty)$, and range $[-1, 1]$

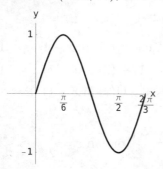

26. Amplitude 3, period π, phase shift 0, domain $(-\infty, \infty)$, and range $[-3, 3]$

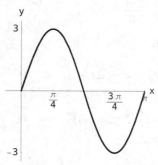

27. Rewriting, we find

$$y = 2\cos(\pi(x - 1)).$$

Thus, we have the following: amplitude 2, period $2\pi/\pi$ or 2, phase shift 1, domain $(-\infty, \infty)$, and range $[-2, 2]$.

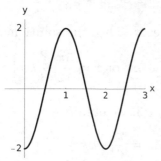

28. Rewriting, we find

$$y = \cos\left(2\left(x - \frac{\pi}{4}\right)\right) + 1.$$

Then we have the following: amplitude 1, period $2\pi/2$ or π, phase shift $\pi/4$, domain $(-\infty, \infty)$, and range is $[0, 2]$.

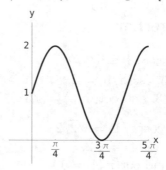

29. The period of

$$y = \tan(x - \pi/2)$$

is π and the phase shift is $\pi/2$. If

$$\cos(x - \pi/2) = 0$$

then

$$x - \frac{\pi}{2} = \frac{\pi}{2} + k\pi$$

or equivalently,

$$x = k\pi$$

where k is an integer. Thus, the domain is

$$\{x : x \neq k\pi\}$$

and the range is $(-\infty, \infty)$.

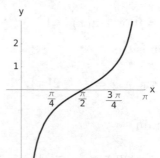

30. The period of

$$y = \tan(\pi x) - 3$$

is π/π or 1, and the phase shift is 0. If

$$\cos(\pi x) = 0$$

then

$$\pi x = \frac{\pi}{2} + k\pi \quad \text{or} \quad x = \frac{1}{2} + k$$

where k is an integer. Thus, the domain is

$$\{x : x \neq \frac{1}{2} + k\}$$

and the range is $(-\infty, \infty)$.

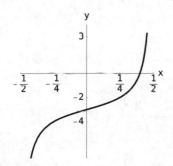

31. opposite, hypotenuse

32. adjacent, hypotenuse

33. one

34. period

35. Pythagorean

36. even, odd

37. oblique

38. law of sines

39. law of cosines

40. triangle inequality

For Thought

1. True, since $i \cdot (-i) = 1$.

2. True, since $\overline{0 + i} = 0 - i = -i$.

3. False, the set of real numbers is a subset of the complex numbers.

4. True, $(\sqrt{3} - i\sqrt{2})(\sqrt{3} + i\sqrt{2}) = 3 + 2 = 5$.

5. True, since $(2 + 5i)(2 - 5i) =$
$2^2 - (5i)^2 = 4 - (-25) = 4 + 25$.

6. False, $5 - \sqrt{-9} = 5 - 3i$.

7. True, since $P(3i) = (3i)^2 + 9 = -9 + 9 = 0$.

8. True, since $(-3i)^2 + 9 = -9 + 9 = 0$.

9. True, since $i^4 = i^2 \cdot i^2 = (-1)(-1) = 1$.

10. False, $i^{18} = (i^4)^4 i^2 = (1)^4(-1) = -1$.

6.1 Exercises

1. complex numbers

3. imaginary number

5. $0 + 6i$, imaginary

7. $\dfrac{1}{3} + \dfrac{1}{3}i$, imaginary

9. $\sqrt{7} + 0i$, real

11. $\dfrac{\pi}{2} + 0i$, real

13. $7 + 2i$

15. $1 - i - 3 - 2i = -2 - 3i$

17. $-18i + 12i^2 = -12 - 18i$

19. $8 + 12i - 12i - 18i^2 = 26$

21. $(5 - 2i)(5 + 2i) = 25 - 4i^2 = 25 - 4(-1) = 29$

23. $(\sqrt{3} - i)(\sqrt{3} + i) = 3 - i^2 = 3 - (-1) = 4$

25. $9 + 24i + 16i^2 = -7 + 24i$

27. $5 - 4i\sqrt{5} + 4i^2 = 1 - 4\sqrt{5}i$

29. $(i^4)^4 \cdot i = (1)^4 \cdot i = i$

31. $(i^4)^{24} i^2 = 1^{24}(-1) = -1$

33. $(i^4)^{-1} = 1^{-1} = 1$

35. Since $i^4 = 1$, we get $i^{-1} = i^{-1}i^4 = i^3 = -i$.

37. $(3 - 9i)(3 + 9i) = 9 - 81i^2 = 90$

39. $\left(\dfrac{1}{2} + 2i\right)\left(\dfrac{1}{2} - 2i\right) = \dfrac{1}{4} - 4i^2 = \dfrac{1}{4} + 4 = \dfrac{17}{4}$

41. $i(-i) = -i^2 = 1$

43. $(3 - i\sqrt{3})(3 + i\sqrt{3}) = 9 - 3i^2 = 9 - 3(-1) = 12$

45. $\dfrac{1}{2 - i} \cdot \dfrac{2 + i}{2 + i} = \dfrac{2 + i}{5} = \dfrac{2}{5} + \dfrac{1}{5}i$

47. $\dfrac{-3i}{1 - i} \cdot \dfrac{1 + i}{1 + i} = \dfrac{-3i + 3}{2} = \dfrac{3}{2} - \dfrac{3}{2}i$

49. $\dfrac{-2 + 6i}{2} = \dfrac{-2}{2} + \dfrac{6i}{2} = -1 + 3i$

51.
$$\dfrac{-3 + 3i}{i} \cdot \dfrac{-i}{-i} = \dfrac{3i - 3i^2}{1} = 3i - 3(-1) = 3 + 3i$$

53.
$$\dfrac{1 - i}{3 + 2i} \cdot \dfrac{3 - 2i}{3 - 2i} = \dfrac{3 - 5i - 2}{13} = \dfrac{1}{13} - \dfrac{5}{13}i$$

55.
$$\dfrac{\sqrt{2} - i\sqrt{3}}{\sqrt{3} + i\sqrt{2}} \cdot \dfrac{\sqrt{3} - i\sqrt{2}}{\sqrt{3} - i\sqrt{2}} = \dfrac{-5i}{5} = -i$$

57. $2i - 3i = -i$

59. $-4 + 2i$

61. $\left(i\sqrt{6}\right)^2 = -6$

63. $(i\sqrt{2})(i\sqrt{50}) = i^2\sqrt{2 \cdot 50}\sqrt{2} = (-1)(2)(5) = -10$

65.
$$\dfrac{-2}{2} + \dfrac{i\sqrt{20}}{2} = -1 + i\dfrac{2\sqrt{5}}{2} = -1 + i\sqrt{5}$$

67. $-3 + \sqrt{9 - 20} = -3 + i\sqrt{11}$

69. $2i\sqrt{2}\left(i\sqrt{2} + 2\sqrt{2}\right) = 4i^2 + 8i = -4 + 8i$

71. $(2i + 2\sqrt{2})(i - \sqrt{2}) = 2(i + \sqrt{2})(i - \sqrt{2}) =$
$2(i^2 - \sqrt{2}^2) = 2(-1 - 2) = -6$

73. $(-2 + i)^2 + 4(-2 + i) + 5 =$
$(3 - 4i) + (-8 + 4i) + 5 = 0$

75. $(-2-i)^2 + 4(-2-i) + 5 =$
$(3+4i) + (-8-4i) + 5 = 0$

77. $(1+i)^2 + 4(1+i) + 5 =$
$(2i) + (4+4i) + 5 = 9 + 6i$

79. $\left(3+i\sqrt{5}\right)^2 - 6(3+i\sqrt{5}) + 14 =$
$\left(4+6i\sqrt{5}\right) - 4 - 6i\sqrt{5} = 0$

81. $(-1)^2 - 6(-1) + 14 = 1 + 6 + 14 = 21$

83. Yes, since $(2i)^2 + 4 = -4 + 4 = 0$

85. No, since $(1-i)^2 + 2(1-i) - 2 = -2i - 2i = -4i$

87. Yes, since $(3-2i)^2 - 6(3-2i) + 13 =$
$5 - 12i - 5 + 12i = 0$

89. Yes, since $3\left(\dfrac{i\sqrt{3}}{3}\right)^2 + 1 = 3\left(\dfrac{-3}{9}\right) + 1 = 0$

91. Yes, $\left(2+i\sqrt{3}\right)^2 - 4(2+i\sqrt{3}) + 7 =$
$\left(1+4i\sqrt{3}\right) - 1 - 4i\sqrt{3} = 0$

93. If r is the remainder when n is divided by 4, then $i^n = i^r$. The possible values of r are $0, 1, 2, 3$ and for i^r they are $1, i, -1, -i$, respectively.

95. Yes, since the product of $a + bi$ with its conjugate is $(a+bi)(a-bi) = a^2 + b^2$, which is a real number.

97. Draw a vector pointing vertically down with a magnitude of 600 lb. Let x be the required force needed to push a 600 lb motorcycle up a ramp with angle of inclination $20°$.

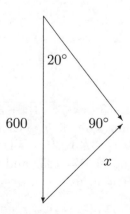

Using right triangle trigonometry, we find

$$\frac{x}{600} = \sin 20°$$

$$x = 600 \sin 20°$$

$$x \approx 205.2 \text{ lb}$$

99. Applying the cosine law, we find

$$\gamma = \cos^{-1}\left(\frac{a^2 + b^2 - c^2}{2ab}\right)$$

$$\gamma = \cos^{-1}\left(\frac{-38}{90}\right)$$

$$\gamma \approx 115.0°.$$

By the sine law, we obtain

$$\beta = \sin^{-1}\left(\frac{9\sin\gamma}{12}\right)$$

$$\beta \approx 42.8°.$$

Then $\alpha = 180° - \gamma - \beta \approx 22.2°$.

101. Simplify as follows:

$$\frac{1 - 2\sin^2 x}{\sin^2 x} - \csc^2 x = \csc^2 x - 2 - \csc^2 x = -2$$

103. Using similar triangles, we find

$$\frac{7}{d} = \frac{x}{1}.$$

By the Pythagorean Theorem, we obtain

$$1 + (d-1)^2 = (7-x)^2.$$

Then we obtain

$$1 + (d-1)^2 = 49\left(1 - \frac{1}{d}\right)^2.$$

Using a computer algebra system, we find

$$d = \frac{\sqrt{47 - 10\sqrt{2}} + 5\sqrt{2} + 1}{2} \approx 6.9016 \text{ ft.}$$

For Thought

1. True, since $(3, -4)$ lies in quadrant 4.

2. False, the absolute value is

$$\sqrt{(-2)^2 + (-5)^2} = \sqrt{29}.$$

3. True, since $r = \sqrt{1^1 + (-3)^2} = \sqrt{10}$
and $\cos\theta = x/r = 1/\sqrt{10}$.

4. False, $\tan\theta = 3/2$.

5. False, $i = 1(\cos 90° + i\sin 90°)$.

6. True, since $\cos 30° = \sqrt{3}/2$ and $30°$ lies
in quadrant 1.

7. True, since $|2 - 5i| = \sqrt{2^2 + (-5)^2} = \sqrt{29}$.

8. True, since $2 - 4i$ lies in quadrant 4
and $\cos\theta = x/r = 2/\sqrt{20} = 1/\sqrt{5}$.

9. True, since $\dfrac{\pi}{4} + \dfrac{\pi}{2} = \dfrac{3\pi}{4}$.

10. False, since
$$\frac{3(\cos\pi/4 + i\sin\pi/4)}{3(\cos\pi/2 + i\sin\pi/2)} =$$
$$1.5(\cos(-\pi/4) + i\sin(-\pi/4)) =$$
$$1.5(\cos\pi/4 - i\sin\pi/4) =$$

6.2 Exercises

1. complex

3. absolute value

5. modulus, argument

7. quotient

9. $\sqrt{2^2 + (-6)^2} = \sqrt{40} = 2\sqrt{10}$

11. $\sqrt{(-2)^2 + (2\sqrt{3})^2} = \sqrt{4 + 12} = 4$

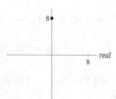

13. $\sqrt{0^2 + 8^2} = 8$

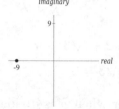

15. $\sqrt{(-9)^2 + 0^2} = 9$

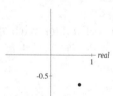

17. $\sqrt{(1/\sqrt{2})^2 + (-1/\sqrt{2})^2} = \sqrt{1/2 + 1/2} = 1$

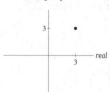

19. $\sqrt{3^2 + 3^2} = \sqrt{18} = 3\sqrt{2}$

21. Since $|-3 + 3i| = \sqrt{(-3)^2 + 3^2} = 3\sqrt{2}$ and if
the terminal side of θ goes through $(-3, 3)$
then $\cos\theta = -3/(3\sqrt{2}) = -1/\sqrt{2}$.
One can choose $\theta = 135°$ and the
trigonometric form is

$$3\sqrt{2}\left(\cos 135° + i\sin 135°\right).$$

23. Since $\sqrt{(-3/\sqrt{2})^2 + (3/\sqrt{2})^2} = 3$ and if the terminal side of θ goes through $(-3/\sqrt{2}, 3/\sqrt{2})$ then $\cos\theta = (-3/\sqrt{2})/3 = -1/\sqrt{2}$. One can choose $\theta = 135°$. The trigonometric form is

$$3\left(\cos 135° + i\sin 135°\right).$$

25. Since terminal side of $0°$ goes through $(8,0)$, the trigonomeric form is

$$8\left(\cos 0° + i\sin 0°\right).$$

27. Since terminal side of $90°$ goes through $(0, \sqrt{3})$, the trigonomeric form is

$$\sqrt{3}\left(\cos 90° + i\sin 90°\right).$$

29. Since $|-\sqrt{3}+i| = \sqrt{(-\sqrt{3})^2 + 1^2} = 2$ and if the terminal side of θ goes through $(-\sqrt{3}, 1)$ then $\cos\theta = -\sqrt{3}/2$. One can choose $\theta = 150°$. The trigonometric form is

$$2\left(\cos 150° + i\sin 150°\right).$$

31. Since $|3+4i| = \sqrt{3^2 + 4^2} = 5$ and if the terminal side of θ goes through $(3, 4)$ then $\cos\theta = 3/5$. One can choose $\theta = \cos^{-1}(3/5) \approx 53.1°$. The trigonometric form is

$$5\left(\cos 53.1° + i\sin 53.1°\right).$$

33. Since $|-3+5i| = \sqrt{(-3)^2 + 5^2} = \sqrt{34}$ and if the terminal side of θ goes through $(-3, 5)$ then $\cos\theta = -3/\sqrt{34}$. One can choose $\theta = \cos^{-1}(-3/\sqrt{34}) \approx 121.0°$. The trigonometric form is

$$\sqrt{34}\left(\cos 121.0° + i\sin 121.0°\right).$$

35. Note $|3-6i| = \sqrt{3^2 + (-6)^2} = \sqrt{45} = 3\sqrt{5}$. If the terminal side of θ goes through $(3, -6)$ then $\tan\theta = -6/3 = -2$. Since $\tan^{-1}(-2) \approx -63.4°$, one can choose $\theta = 360° - 63.4° = 296.6°$. The trigonometric form is

$$3\sqrt{5}\left(\cos 296.6° + i\sin 296.6°\right).$$

37. $\sqrt{2}\left(\dfrac{1}{\sqrt{2}} + i\dfrac{1}{\sqrt{2}}\right) = 1 + i$

39. $\dfrac{\sqrt{3}}{2}\left(-\dfrac{\sqrt{3}}{2} + i\dfrac{1}{2}\right) = -\dfrac{3}{4} + i\dfrac{\sqrt{3}}{4}$

41. $\dfrac{1}{2}(-0.848 - 0.53i) \approx -0.42 - 0.26i$

43. $3(0 + i) = 3i$

45. $\sqrt{3}(0 - i) = -i\sqrt{3}$

47. $\sqrt{6}\left(\dfrac{1}{2} + i\dfrac{\sqrt{3}}{2}\right) =$

$\dfrac{\sqrt{6}}{2} + i\dfrac{\sqrt{18}}{2} = \dfrac{\sqrt{6}}{2} + \dfrac{3\sqrt{2}}{2}i$

49. $6\left(\cos 450° + i\sin 450°\right) =$
$6\left(\cos 90° + i\sin 90°\right) = 6(0 + i) = 6i$

51. $\sqrt{6}\left(\cos 30° + i\sin 30°\right) =$

$\sqrt{6}\left(\dfrac{\sqrt{3}}{2} + i\dfrac{1}{2}\right) =$

$\dfrac{\sqrt{18}}{2} + i\dfrac{\sqrt{6}}{2} = \dfrac{3\sqrt{2}}{2} + \dfrac{\sqrt{6}}{2}i$

53. $9\left(\cos 90° + i\sin 90°\right) =$
$9(0 + i) = 9i$

55. $2\left(\cos(\pi/6) + i\sin(\pi/6)\right) =$

$2\left(\dfrac{\sqrt{3}}{2} + i\dfrac{1}{2}\right) = \sqrt{3} + i$

57. $0.5\left(\cos(-47.5°) + i\sin(-47.5°)\right) \approx$
$0.5\left(0.6756 - i \cdot 0.7373\right) \approx 0.34 - 0.37i$

59. Since $z_1 = 4\sqrt{2}\left(\cos 45° + i\sin 45°\right)$ and $z_2 = 5\sqrt{2}\left(\cos 225° + i\sin 225°\right)$, we have $z_1 z_2 = 40\left(\cos 270° + i\sin 270°\right) =$
$40(0 - i) = -40i$ and
$\dfrac{z_1}{z_2} = 0.8\left(\cos(-180°) + i\sin(-180°)\right) =$
$0.8(-1 + i \cdot 0) = -0.8$

61. Since $z_1 = 2\left(\cos 30° + i\sin 30°\right)$ and $z_2 = 4\left(\cos 60° + i\sin 60°\right)$, we have $z_1 z_2 = 8\left(\cos 90° + i\sin 90°\right) =$
$8(0 + i) = 8i$ and
$\dfrac{z_1}{z_2} = \dfrac{1}{2}\left(\cos(-30°) + i\sin(-30°)\right) =$
$\dfrac{1}{2}\left(\dfrac{\sqrt{3}}{2} - i\dfrac{1}{2}\right) = \dfrac{\sqrt{3}}{4} - \dfrac{1}{4}i.$

63. Since $z_1 = 2\sqrt{2}\,(\cos 45° + i\sin 45°)$ and
$z_2 = 2\,(\cos 315° + i\sin 315°)$, we have
$z_1 z_2 = 4\sqrt{2}\,(\cos 360° + i\sin 360°) =$
$4\sqrt{2}\,(1 + i \cdot 0) = 4\sqrt{2}$ and
$\dfrac{z_1}{z_2} = \sqrt{2}\,(\cos(-270°) + i\sin(-270°)) =$
$\sqrt{2}\,(0 + i) = i\sqrt{2}.$

65. Let α and β be angles whose terminal side
goes through $(3, 4)$ and $(-5, -2)$, respectively.
Since $|3 + 4i| = 5$ and $|-5 - 2i| = \sqrt{29}$, we
have $\cos\alpha = 3/5$, $\sin\alpha = 4/5$,
$\cos\beta = -5/\sqrt{29}$, and $\sin\beta = -2/\sqrt{29}$. From
the sum and difference identities, we find
$$\cos(\alpha + \beta) = -\frac{7}{5\sqrt{29}},$$
$$\sin(\alpha + \beta) = -\frac{26}{5\sqrt{29}},$$
$$\cos(\alpha - \beta) = -\frac{23}{5\sqrt{29}},$$
$$\sin(\alpha - \beta) = -\frac{14}{5\sqrt{29}}.$$
Note $z_1 = 5(\cos\alpha + i\sin\alpha)$ and
$z_2 = \sqrt{29}(\cos\beta + i\sin\beta)$. Then
$$\begin{aligned}
z_1 z_2 &= 5\sqrt{29}\,(\cos(\alpha + \beta) + i\sin(\alpha + \beta)) \\
&= 5\sqrt{29}\left(-\frac{7}{5\sqrt{29}} - i\frac{26}{5\sqrt{29}}\right) \\
z_1 z_2 &= -7 - 26i
\end{aligned}$$
and
$$\begin{aligned}
\frac{z_1}{z_2} &= \frac{5}{\sqrt{29}}\,(\cos(\alpha - \beta) + i\sin(\alpha - \beta)) \\
&= \frac{5}{\sqrt{29}}\left(-\frac{23}{5\sqrt{29}} - i\frac{14}{5\sqrt{29}}\right) \\
\frac{z_1}{z_2} &= -\frac{23}{29} - \frac{14}{29}i.
\end{aligned}$$

67. Let α and β be angles whose terminal sides
go through $(2, -6)$ and $(-3, -2)$,
respectively. Since $|2 - 6i| = 2\sqrt{10}$ and
$|-3 - 2i| = \sqrt{13}$, we have $\cos\alpha = 1/\sqrt{10}$,
$\sin\alpha = -3/\sqrt{10}$, $\cos\beta = -3/\sqrt{13}$, and
$\sin\beta = -2/\sqrt{13}$. By using the sum and
difference identities, we obtain
$$\cos(\alpha + \beta) = -\frac{9}{\sqrt{130}},$$

$$\sin(\alpha + \beta) = \frac{7}{\sqrt{130}},$$
$$\cos(\alpha - \beta) = \frac{3}{\sqrt{130}},$$
$$\sin(\alpha - \beta) = \frac{11}{\sqrt{130}}.$$
Note $z_1 = 2\sqrt{10}(\cos\alpha + i\sin\alpha)$ and
$z_2 = \sqrt{13}(\cos\beta + i\sin\beta)$. Thus,
$$\begin{aligned}
z_1 z_2 &= 2\sqrt{130}\,(\cos(\alpha + \beta) + i\sin(\alpha + \beta)) \\
&= 2\sqrt{130}\left(-\frac{9}{\sqrt{130}} + i\frac{7}{\sqrt{130}}\right) \\
z_1 z_2 &= -18 + 14i
\end{aligned}$$
and
$$\begin{aligned}
\frac{z_1}{z_2} &= \frac{2\sqrt{10}}{\sqrt{13}}\,(\cos(\alpha - \beta) + i\sin(\alpha - \beta)) \\
&= \frac{2\sqrt{10}}{\sqrt{13}}\left(\frac{3}{\sqrt{130}} + i\frac{11}{\sqrt{130}}\right) \\
\frac{z_1}{z_2} &= \frac{6}{13} + \frac{22}{13}i.
\end{aligned}$$

69. Note $3i = 3(\cos 90° + i\sin 90°)$ and
$1 + i = \sqrt{2}(\cos 45° + i\sin 45°)$. Then we get
$$\begin{aligned}
(3i)(1 + i) &= 3\sqrt{2}\,(\cos 135° + i\sin 135°) \\
&= 3\sqrt{2}\left(-\frac{\sqrt{2}}{2} + i\cdot\frac{\sqrt{2}}{2}\right) \\
(3i)(1 + i) &= -3 + 3i
\end{aligned}$$
and
$$\begin{aligned}
\frac{3i}{1 + i} &= \frac{3}{\sqrt{2}}\,(\cos 45° + i\sin 45°) \\
&= \frac{3}{\sqrt{2}}\left(\frac{\sqrt{2}}{2} + i\cdot\frac{\sqrt{2}}{2}\right) \\
(3i)(1 + i) &= 1.5 + 1.5i.
\end{aligned}$$

71. $\left[3\left(\cos\dfrac{\pi}{6} + i\sin\dfrac{\pi}{6}\right)\right]\left[3\left(\cos\dfrac{\pi}{6} - i\sin\dfrac{\pi}{6}\right)\right]$
$$= 9\left(\cos^2\frac{\pi}{6} + \sin^2\frac{\pi}{6}\right) = 9(1) = 9$$

73. $[2(\cos 7° + i\sin 7°)]\,[2(\cos 7° - i\sin 7°)]$
$= 4(\cos^2 7° + \sin^2 7°) = 4(1) = 4$

75. Since $3 + 3i = 3\sqrt{2}\left(\cos 45° + i\sin 45°\right)$, we get
$$(3 + 3i)^3 =$$
$$(3\sqrt{2})^3 \left(\cos(3 \cdot 45°) + i\sin(3 \cdot 45°)\right) =$$
$$54\sqrt{2}\left(\cos 135° + i\sin 135°\right) =$$
$$54\sqrt{2}\left(-\frac{1}{\sqrt{2}} + i\frac{1}{\sqrt{2}}\right) = -54 + 54i.$$

77. The reciprocal of z is $\dfrac{1}{z} = \dfrac{\cos 0 + i\sin 0}{r\left[\cos\theta + i\sin\theta\right]} =$
$$r^{-1}\left[\cos(0 - \theta) + i\sin(0 - \theta)\right] =$$
$$r^{-1}\left[\cos\theta - i\sin\theta\right] \text{ provided } r \neq 0.$$

79. Using polar form, we find
$$6\left(\cos 9° + i\sin 9°\right) + 3\left(\cos 5° + i\sin 5°\right) =$$
$$6\cos 9° + 3\cos 5° + i\left(6\sin 9° + 3\sin 5°\right).$$

Since
$$(1 + 3i) + (5 - 7i) = 6 - 4i$$

it is easier to add complex numbers in standard form.

81. **a)** $9 + 30i + 25i^2 = -16 + 30i$

b) $\dfrac{1 + i}{2 + i} \cdot \dfrac{2 - i}{2 - i} = \dfrac{2 - i + 2i + 1}{5} = \dfrac{3}{5} + \dfrac{1}{5}i$

c) $2\sqrt{2}i + 5\sqrt{2}i = 7\sqrt{2}i$

83. Note, the maximum and minimum y-values of the given key points are -1 and -3, respectively. Since one-half of the difference between the maximum and minimum y-values is 1, we conclude $A = 1$.

Since the midpoint of the maximum and minimum y-values is -2, we find $D = -2$.

Since the first key point is $(\pi/3, -1)$, the phase shift is $C = \frac{\pi}{3}$.

Since the difference between the last and first x-coordinates is the period, we obtain
$$\frac{2\pi}{B} = \frac{5\pi}{3} - \frac{\pi}{3} = \frac{4\pi}{3}.$$

Then $B = \frac{3}{2} = 1.5$ Thus, an equation is
$$y = \cos\left(1.5\left(x - \frac{\pi}{3}\right)\right) - 2.$$

85. **a)** $\dfrac{2\pi}{B} = \dfrac{2\pi}{2} = \pi$

b) $\dfrac{2\pi}{B} = \dfrac{2\pi}{2\pi} = 1$

c) $\dfrac{\pi}{B} = \dfrac{\pi}{4}$

d) $\dfrac{2\pi}{B} = \dfrac{2\pi}{3}$

87. Suppose there are n rows of circles with n circles in the bottom row, $n-1$ circles in the next, ..., and one in the top row. Then the number of circles is
$$1 + 2 + 3 + \dots + n = \frac{n(n+1)}{2}.$$

a) Look at the n circles in the bottom row. The distance between the points of tangency to the bottom side of the equilateral triangle of the first and last circle is $2r(n-1)$. Also, the distance from the last point of tangency to the closest vertex of the triangle is $r\sqrt{3}$. Since the side of the triangle has length 1, we have
$$2r(n - 1) + 2r\sqrt{3} = 1.$$

Solving for r, we obtain
$$r = \frac{1}{2(n + \sqrt{3} - 1)}.$$

Then the sum of the areas of all the circles is
$$A(n) = \frac{n(n+1)}{2}\pi r^2$$
$$= \frac{n(n+1)}{2}\pi\left(\frac{1}{2(n + \sqrt{3} - 1)}\right)^2$$
$$A(n) = \frac{\pi(n^2 + n)}{8(n + \sqrt{3} - 1)^2}.$$

b) By dividing the leading coefficients of the numerator and denominator of the rational function $A(n)$, we find that $A(n)$ approaches
$$\frac{\pi}{8} \approx 0.39$$

as n approaches infinity.

Since the area of the equilateral triangle is
$$\frac{\sqrt{3}}{4} \approx 0.43$$

the triangular pipe will not be filled by the circular cables.

For Thought

1. False, $(2 + 3i)^2 = 4 + 12i + 9i^2$.

2. False, $z^3 = 8(\cos 360° + i \sin 360°) = 8$.

3. True   4.   False, the argument is 4θ.

5. False, since $\left[\dfrac{1}{2} + i \cdot \dfrac{1}{2}\right]^2 = \dfrac{i}{2}$ and
   $\cos 2\pi/3 + i \sin \pi/3 = -\dfrac{1}{2} + i \cdot \dfrac{\sqrt{3}}{2}$.

6. False, since $\cos 5\pi/6 = \cos 7\pi/6$ and
   $5\pi/6 \neq 7\pi/6 + 2k\pi$ for any integer k.
   It is possible that $\alpha = 2k\pi - \beta$.

7. True, since $360/5 = 72$.   8.   True, since $|x| = 1$.

9. True, $x = \cos(45°k) + i \sin(45°k)$ where k is an integer.

10. False, there are four imaginary solutions.

6.3 Exercises

1. De Moivre's

3. $3^3(\cos 90° + i \sin 90°) = 27(0 + i) = 27i$

5. $(\sqrt{2})^4(\cos 480° + i \sin 480°) =$
   $4(\cos 120° + i \sin 120°) =$
   $4\left(-\dfrac{1}{2} + i \cdot \dfrac{\sqrt{3}}{2}\right) = -2 + 2i\sqrt{3}$

7. $\cos(8\pi/12) + i \sin(8\pi/12) =$
   $\cos(2\pi/3) + i \sin(2\pi/3) = -\dfrac{1}{2} + i\dfrac{\sqrt{3}}{2}$

9. $(\sqrt{6})^4 [\cos(8\pi/3) + i \sin(8\pi/3)] =$
   $36 [\cos(2\pi/3) + i \sin(2\pi/3)] =$
   $36\left[-\dfrac{1}{2} + i\dfrac{\sqrt{3}}{2}\right] = -18 + 18i\sqrt{3}$

11. $4.3^5 [\cos 61.5° + i \sin 61.5°] \approx$
    $1470.1 [0.4772 + 0.8788i] \approx 701.5 + 1291.9i$

13. $\left(2\sqrt{2} [\cos 45° + i \sin 45°]\right)^3 =$
    $16\sqrt{2} [\cos 135° + i \sin 135°] =$
    $16\sqrt{2}\left[-\dfrac{1}{\sqrt{2}} + i\dfrac{1}{\sqrt{2}}\right] = -16 + 16i$

15. $(2 [\cos(-30°) + i \sin(-30°)])^4 =$
    $16 [\cos(-120°) + i \sin(-120°)] =$
    $16\left[-\dfrac{1}{2} - i\dfrac{\sqrt{3}}{2}\right] = -8 - 8i\sqrt{3}$

17. $(6 [\cos 240° + i \sin 240°])^5 =$
    $7776 [\cos 1200° + i \sin 1200°] =$
    $7776 [\cos 120° + i \sin 120°] =$
    $7776\left[-\dfrac{1}{2} + i\dfrac{\sqrt{3}}{2}\right] = -3888 + 3888i\sqrt{3}$

19. Note $|2 + 3i| = \sqrt{13}$. If the terminal side of α goes through $(2, 3)$ then $\cos\alpha = 2/\sqrt{13}$ and $\sin\alpha = 3/\sqrt{13}$. By using the double-angle identities one can successively obtain
    $\cos 2\alpha = -5/13,\ \sin 2\alpha = 12/13,$
    $\cos 4\alpha = -119/169,\ \sin 4\alpha = -120/169.$
    So $(2 + 3i)^4 = \left(\sqrt{13} [\cos\alpha + i \sin\alpha]\right)^4 =$
    $169(\cos 4\alpha + i \sin 4\alpha) =$
    $169 (-119/169 - 120i/169) = -119 - 120i$

21. Note $|2 - i| = \sqrt{5}$. If the terminal side of α goes through $(2, -1)$ then $\cos\alpha = 2/\sqrt{5}$ and $\sin\alpha = -1/\sqrt{5}$. By using the double-angle identities one can successively obtain
    $\cos 2\alpha = 3/5,\ \sin 2\alpha = -4/5,$
    $\cos 4\alpha = -7/25,\ \sin 4\alpha = -24/25.$
    So $(2 - i)^4 = \left(\sqrt{5} [\cos\alpha + i \sin\alpha]\right)^4 =$
    $25(\cos 4\alpha + i \sin 4\alpha) =$
    $25 (-7/25 - 24i/25) = -7 - 24i.$

23. Let $\omega = |1.2 + 3.6i|$. If the terminal side of α goes through $(1.2, 3.6)$ then $\cos\alpha = 1.2/\omega$ and $\sin\alpha = 3.6/\omega$. By using the double-angle identities, one obtains
    $\cos 2\alpha = -11.52/\omega^2$ and $\sin 2\alpha = 8.64/\omega^2$.
    By the sum identities, one gets
    $\cos 3\alpha = \cos(2\alpha + \alpha) = -44.928/\omega^3$ and
    $\sin 3\alpha = \sin(2\alpha + \alpha) = -31.104/\omega^3$.
    So $(1.2 + 3.6i)^3 = (\omega [\cos\alpha + i \sin\alpha])^3 =$
    $\omega^3(\cos 3\alpha + i \sin 3\alpha) =$
    $\omega^3 (-44.928/\omega^3 - 31.104i/\omega^3) =$
    $-44.928 - 31.104i.$

25. The square roots are given by
    $2\left[\cos\left(\dfrac{90° + k360°}{2}\right) + i \sin\left(\dfrac{90° + k360°}{2}\right)\right] =$
    $2 [\cos(45° + k \cdot 180°) + i \sin(45° + k \cdot 180°)]$

where k is an integer. If $k = 0, 1$, we get

$2\left(\cos 45° + i \sin 45°\right)$ and

$2\left(\cos 225° + i \sin 225°\right).$

27. The fourth roots are given by

$$\cos\left(\frac{120° + k360°}{4}\right) + i\sin\left(\frac{120° + k360°}{4}\right) =$$

$$\cos(30° + k \cdot 90°) + i\sin(30° + k \cdot 90°)$$

where k is an integer. If $k = 0, 1, 2, 3$, we find

$$\cos\alpha + i\sin\alpha$$

where $\alpha = 30°, 120°, 210°, 300°$.

29. The sixth roots are

$$2\left[\cos\left(\frac{\pi + 2k\pi}{6}\right) + i\sin\left(\frac{\pi + 2k\pi}{6}\right)\right]$$

where k is an integer. If $k = 0, 1, 2, 3, 4, 5$, we have

$$2\left(\cos\alpha + i\sin\alpha\right)$$

where $\alpha = \frac{\pi}{6}, \frac{\pi}{2}, \frac{5\pi}{6}, \frac{7\pi}{6}, \frac{3\pi}{2}, \frac{11\pi}{6}$.

31. The cube roots of 1 are given by

$$\cos\left(\frac{k360°}{3}\right) + i\sin\left(\frac{k360°}{3}\right)$$

where k is an integer. If $k = 0, 1, 2$, we get

$\cos 0° + i\sin 0° = 1,$

$\cos 120° + i\sin 120° = -\frac{1}{2} + i\frac{\sqrt{3}}{2}$, and

$\cos 240° + i\sin 240° = -\frac{1}{2} - i\frac{\sqrt{3}}{2}.$

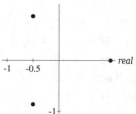

33. The fourth roots of 16 are

$$2\left[\cos\left(\frac{k360°}{4}\right) + i\sin\left(\frac{k360°}{4}\right)\right]$$

where k is an integer. If $k = 0, 1, 2, 3$, we obtain

$2\left[\cos 0° + i\sin 0°\right] = 2,$

$2\left[\cos 90° + i\sin 90°\right] = 2i,$

$2\left[\cos 180° + i\sin 180°\right] = -2$, and

$2\left[\cos 270° + i\sin 270°\right] = -2i.$

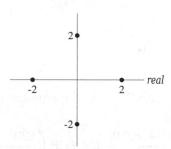

35. The fourth roots of -1 are

$$\cos\left(\frac{180° + k360°}{4}\right) + i\sin\left(\frac{180° + k360°}{4}\right) =$$

$$\cos\left(45° + k90°\right) + i\sin\left(45° + k90°\right)$$

where k is an integer. If $k = 0, 1, 2, 3$, we find

$\cos 45° + i\sin 45° = \frac{\sqrt{2}}{2} + \frac{\sqrt{2}}{2}i,$

$\cos 135° + i\sin 135° = -\frac{\sqrt{2}}{2} + \frac{\sqrt{2}}{2}i,$

$\cos 225° + i\sin 225° = -\frac{\sqrt{2}}{2} - \frac{\sqrt{2}}{2}i$, and

$\cos 315° + i\sin 315° = \frac{\sqrt{2}}{2} - \frac{\sqrt{2}}{2}i.$

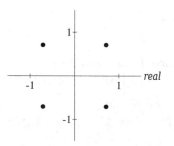

37. The cube roots of i are

$$\cos\left(\frac{90° + k360°}{3}\right) + i\sin\left(\frac{90° + k360°}{3}\right) =$$

$$\cos\left(30° + k120°\right) + i\sin\left(30° + k120°\right)$$

where k is an integer. If $k = 0, 1, 2$, we have

$\cos 30° + i\sin 30° = \frac{\sqrt{3}}{2} + \frac{1}{2}i,$

$\cos 150° + i\sin 150° = -\frac{\sqrt{3}}{2} + \frac{1}{2}i$, and

$\cos 270° + i \sin 270° = -i.$

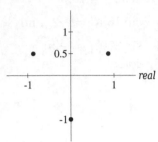

39. Since $|-2 + 2i\sqrt{3}| = 4$, the square roots are given by

$$2 \left[\cos \left(\frac{120° + k360°}{2} \right) + i \sin \left(\frac{120° + k360°}{2} \right) \right]$$
$$= 2 \left[\cos \left(60° + k180° \right) + i \sin \left(60° + k180° \right) \right].$$

If $k = 0, 1$ one obtains
$2 \left[\cos 60° + i \sin 60° \right] = 1 + i\sqrt{3}$, and
$2 \left[\cos 240° + i \sin 240° \right] = -1 - i\sqrt{3}.$

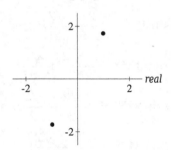

41. Note $|1 + 2i| = \sqrt{5}$. Since $\tan^{-1} 2 \approx 63.4°$ and

$$\frac{63.4° + k360°}{2} = 31.7° + k180°$$

the square roots are given by
$\sqrt[4]{5} \left[\cos \left(31.7° + k180° \right) + i \sin \left(31.7° + k180° \right) \right].$

If $k = 0, 1$ one obtains
$\sqrt[4]{5} \left[\cos 31.7° + i \sin 31.7° \right] \approx 1.272 + 0.786i,$
and
$\sqrt[4]{5} \left[\cos 211.7° + i \sin 211.7° \right] \approx -1.272 - 0.786i.$

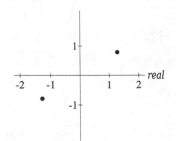

43. Solutions are the cube roots of -1. Namely,

$$\cos \left(\frac{180° + k360°}{3} \right) + i \sin \left(\frac{180° + k360°}{3} \right)$$

$= \cos \left(60° + k120° \right) + i \sin \left(60° + k120° \right).$
If $k = 0, 1, 2$ one obtains
$\cos 60° + i \sin 60° = \frac{1}{2} + \frac{\sqrt{3}}{2}i,$
$\cos 180° + i \sin 180° = -1,$ and
$\cos 300° + i \sin 300° = \frac{1}{2} - \frac{\sqrt{3}}{2}i.$

45. Solutions are the fourth roots of 81. Namely,

$$3 \left[\cos \left(\frac{k360°}{4} \right) + i \sin \left(\frac{k360°}{4} \right) \right]$$
$$= 3 \left[\cos \left(k90° \right) + i \sin \left(k90° \right) \right].$$

If $k = 0, 1, 2, 3$ one obtains
$3 \left[\cos 0° + i \sin 0° \right] = 3,$
$3 \left[\cos 90° + i \sin 90° \right] = 3i,$
$3 \left[\cos 180° + i \sin 180° \right] = -3$ and
$3 \left[\cos 270° + i \sin 270° \right] = -3i.$

47. Solutions are the square roots of $-2i$.

If $k = 0, 1$ in $\alpha = \dfrac{-90° + k360°}{2}$ then

$\alpha = -45°, 135°.$ These roots are
$\sqrt{2} \left[\cos(-45°) + i \sin(-45°) \right] = 1 - i$ and
$\sqrt{2} \left[\cos 135° + i \sin 135° \right] = -1 + i.$

49. Solutions of $x(x^6 - 64) = 0$ are $x = 0$ and the sixth roots of 64. The sixth roots are given by
$$2 \left[\cos \left(\frac{k360°}{6} \right) + i \sin \left(\frac{k360°}{6} \right) \right]$$
$$= 2 \left[\cos \left(k60° \right) + i \sin \left(k60° \right) \right].$$

If $k = 0, 1, 2, 3, 4, 5$ one obtains
$2 \left[\cos 0° + i \sin 0° \right] = 2,$
$2 \left[\cos 60° + i \sin 60° \right] = 1 + i\sqrt{3},$
$2 \left[\cos 120° + i \sin 120° \right] = -1 + i\sqrt{3},$
$2 \left[\cos 180° + i \sin 180° \right] = -2,$
$2 \left[\cos 240° + i \sin 240° \right] = -1 - i\sqrt{3},$ and
$2 \left[\cos 300° + i \sin 300° \right] = 1 - i\sqrt{3}.$

51. Solutions are the fifth roots of 2. Namely,
$$\sqrt[5]{2} \left[\cos \left(\frac{k360°}{5} \right) + i \sin \left(\frac{k360°}{5} \right) \right]$$
$$= \sqrt[5]{2} \left[\cos \left(k72° \right) + i \sin \left(k72° \right) \right].$$

Solutions are $x = \sqrt[5]{2} \left[\cos \alpha + i \sin \alpha \right]$
where $\alpha = 0°, 72°, 144°, 216°, 288°.$

53. Solutions are the fourth roots of $-3 + i$. Since $|-3+i| = \sqrt{10}$, an argument of $-3+i$ is $\cos^{-1}(-3/\sqrt{10}) \approx 161.6°$. Arguments of the fourth roots are given by

$$\frac{161.6° + k360°}{4} = 40.4° + k90°.$$

By choosing $k = 0, 1, 2, 3$, the solutions are $x = \sqrt[8]{10}\,[\cos\alpha + i\sin\alpha]$ where $\alpha = 40.4°, 130.4°, 220.4°, 310.4°$.

55. $[\cos\pi/3 + i\sin\pi/6]^3 =$

$$[1/2 + i(1/2)]^3 = \left[\frac{1}{2}(1+i)\right]^3 =$$

$$\frac{1}{8}\left[\sqrt{2}\,(\cos 45° + i\sin 45°)\right]^3 =$$

$$\frac{1}{8}\left[2\sqrt{2}\,(\cos 135° + i\sin 135°)\right] =$$

$$\frac{\sqrt{2}}{4}\left(-\frac{\sqrt{2}}{2} + i\cdot\frac{\sqrt{2}}{2}\right) = -\frac{1}{4} + \frac{1}{4}i$$

57. By the quadratic formula, we get

$$x = \frac{-(-1+i) \pm \sqrt{(-1+i)^2 - 4(1)(-i)}}{2}$$

$$x = \frac{1-i \pm \sqrt{(1-2i-1)+4i}}{2}$$

$$x = \frac{1-i \pm \sqrt{2i}}{2}.$$

Note the square roots of $2i$ are given by $\sqrt{2}(\cos 45° + i\sin 45°) = 1+i$ and $\sqrt{2}(\cos 225° + i\sin 225°) = -1-i$. These two roots differ by a minus sign. So

$$x = \frac{1-i \pm (1+i)}{2}$$

$$x = \frac{1-i+1+i}{2} \quad \text{or} \quad x = \frac{1-i-1-i}{2}$$

$$x = 1 \quad \text{or} \quad x = -i.$$

The solutions are $x = 1, -i$.

61. $|3+5i| = \sqrt{3^2 + 5^2} = \sqrt{34}$

63. Since $\dfrac{\pi}{6} + \dfrac{\pi}{3} = \dfrac{\pi}{2}$, the product is

$$6\left(\cos\frac{\pi}{2} + i\sin\frac{\pi}{2}\right) = 6i.$$

65. Since $\sin(2\pi x) = \pm\dfrac{\sqrt{2}}{2}$, we find

$$2\pi x = \frac{\pi}{4} + \frac{k\pi}{2}$$

$$x = \frac{1}{8} + \frac{k}{4}$$

Choose $k = 0, 1, 2, 3, 4, 5$ since x lies between 0 and $\dfrac{\pi}{2} \approx 1.57$. Then

$$x = \frac{1}{8}, \frac{3}{8}, \frac{5}{8}, \frac{7}{8}, \frac{9}{8}, \frac{11}{8}.$$

67. Let $\beta = 2\alpha$. By the cosine law, we find

$$a^2 = b^2 + c^2 - 2bc\cos\alpha$$

and

$$b^2 = a^2 + c^2 - 2ac\cos 2\alpha.$$

Since $\cos 2\alpha = 2\cos^2\alpha - 1$, we obtain

$$b^2 = a^2 + c^2 - 2ac\left[2\left(\frac{b^2 + c^2 - a^2}{2bc}\right)^2 - 1\right].$$

We may rewrite the above equation as

$$\frac{(a-b-c)(a+b-c)(a+c)(a^2 - b^2 + ac)}{b^2 c} = 0.$$

Since the sum of any two sides of a triangle must be larger than the remaining side, we obtain

$$a^2 - b^2 + ac = 0$$

or equivalently

$$c = \frac{b^2 - a^2}{a}.$$

a) Using the above equation, such a triangle with the smallest perimeter has sides

$$a = 4, b = 6, c = 5.$$

b) In the beginning of the above discussion, we have $\beta = 2\alpha$ and then obtained

$$c = \frac{b^2 - a^2}{a}.$$

The next two larger such triangles have sides

$$a = 9, b = 12, c = 7$$

and

$$a = 8, b = 12, c = 10.$$

For Thought

1. True, since the distance is $|r|$.

2. False, the distance is $|r|$.

3. False

4. False, $x = r\cos\theta$, $y = r\sin\theta$, and $x^2 + y^2 = r^2$.

5. True, since $x = -4\cos 225° = 2\sqrt{2}$ and $y = -4\sin 225° = 2\sqrt{2}$.

6. True, $\theta = \pi/4$ is a straight line through the origin which makes an angle of $\pi/4$ with the positive x-axis.

7. True, since each circle is centered at the origin with radius 5.

8. False, since upon substitution one gets $\cos 2\pi/3 = -1/2$ while $r^2 = 1/2$.

9. False, for $r = 1/\sin\theta$ is undefined when $\theta = k\pi$ for any integer k.

10. False, since $r = \theta$ is a reflection of $r = -\theta$ about the origin.

6.4 Exercises

1. polar

3. four-leaf rose

5. $r\cos\theta$, $r\sin\theta$

7. Since $r = \sqrt{3^2 + 3^2} = 3\sqrt{2}$ and $\theta = \tan^{-1}(3/3) = \pi/4$, $(r, \theta) = (3\sqrt{2}, \pi/4)$.

9. $(3, \pi/2)$

11. $(3, \pi/6)$

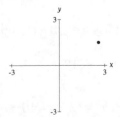

13. $(-2, 2\pi/3)$

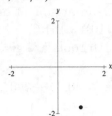

15. $(2, -\pi/4)$

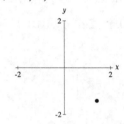

17. $(3, -225°)$

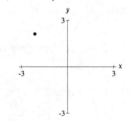

19. $(-2, 45°)$

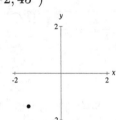

21. $(4, 390°)$

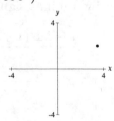

23. $(x, y) = (1 \cdot \cos(\pi/6), 1 \cdot \sin(\pi/6)) = \left(\dfrac{\sqrt{3}}{2}, \dfrac{1}{2}\right)$

25. $(x, y) = (-3\cos(3\pi/2), -3\sin(3\pi/2)) = (0, 3)$

27. $(x, y) = \left(\sqrt{2}\cos 135°, \sqrt{2}\sin 135°\right) = (-1, 1)$

29. $(x, y) = \left(-\sqrt{6}\cos(-60°), -\sqrt{6}\sin(-60°)\right) =$
$\left(-\dfrac{\sqrt{6}}{2}, \dfrac{3\sqrt{2}}{2}\right)$

31. Since $r = \sqrt{(\sqrt{3})^2 + 3^2} = 2\sqrt{3}$ and

$\cos\theta = \dfrac{x}{r} = \dfrac{\sqrt{3}}{2\sqrt{3}} = \dfrac{1}{2}$, one can choose

$\theta = 60°$. Then

$$(r, \theta) = (2\sqrt{3}, 60°).$$

33. Since $r = \sqrt{(-2)^2 + 2^2} = 2\sqrt{2}$ and

$\cos\theta = \dfrac{x}{r} = \dfrac{-2}{2\sqrt{2}} = -\dfrac{1}{\sqrt{2}}$, one can choose

$\theta = 135°$. Then

$$(r, \theta) = (2\sqrt{2}, 135°).$$

35. $(r, \theta) = (2, 90°)$

37. Note $r = \sqrt{(-3)^2 + (-3)^2} = 3\sqrt{2}$.

Since $\tan\theta = \dfrac{y}{x} = \dfrac{-3}{-3} = 1$ and $(-3, -3)$

is in quadrant III, one can choose $\theta = 225°$.
Then
$$(r, \theta) = (3\sqrt{2}, 225°).$$

39. Since $r = \sqrt{1^2 + 4^2} = \sqrt{17}$ and

$\theta = \cos^{-1}\left(\dfrac{x}{r}\right) = \cos^{-1}\left(\dfrac{1}{\sqrt{17}}\right) \approx 75.96°$,

we find
$$(r, \theta) = (\sqrt{17}, 75.96°).$$

41. Since $r = \sqrt{(\sqrt{2})^2 + (-2)^2} = \sqrt{6}$ and

$\theta = \tan^{-1}\left(\dfrac{y}{x}\right) = \tan^{-1}\left(\dfrac{-2}{\sqrt{2}}\right) \approx -54.7°$,

we find
$$(r, \theta) = (\sqrt{6}, -54.7°).$$

43. $r = 2\sin\theta$ is a circle centered at $(x, y) = (0, 1)$.
It goes through the following points in polar
coordinates: $(0, 0)$, $(1, \pi/6)$, $(2, \pi/2)$,
$(1, 5\pi/6)$, $(0, \pi)$.

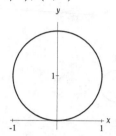

45. $r = 3\cos 2\theta$ is a four-leaf rose that goes
through the following points in polar
coordinates $(3, 0)$, $(3/2, \pi/6)$,
$(0, \pi/4)$, $(-3, \pi/2)$, $(0, 3\pi/4)$, $(3, \pi)$,
$(-3, 3\pi/2)$, $(0, 7\pi/4)$.

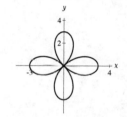

47. $r = 2\theta$ is spiral-shaped and goes through the
following points in polar coordinates
$(-\pi, -\pi/2)$, $(0, 0)$, $(\pi, \pi/2)$, $(2\pi, \pi)$

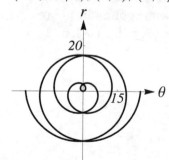

49. $r = 1 + \cos\theta$ goes through the following points
in polar coordinates $(2, 0)$, $(1.5, \pi/3)$, $(1, \pi/2)$,
$(0.5, 2\pi/3)$, $(0, \pi)$.

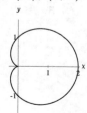

51. $r^2 = 9\cos 2\theta$ goes through the following points in polar coordinates $(0, -\pi/4)$, $(\pm 3\sqrt{2}/2, -\pi/6)$, $(\pm 3, 0)$, $(\pm 3\sqrt{2}/2, \pi/6)$, and $(0, \pi/4)$.

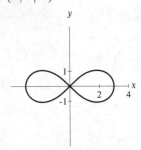

53. $r = 4\cos 2\theta$ is a four-leaf rose that goes through the following points in polar coordinates $(4, 0)$, $(2, \pi/6)$, $(0, \pi/4)$, $(-4, \pi/2)$, $(2, 5\pi/6)$, $(4, \pi)$, $(0, 5\pi/4)$, $(-4, 3\pi/2)$, $(0, 7\pi/4)$.

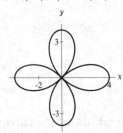

55. $r = 2\sin 3\theta$ is a three-leaf rose that goes through the following points in polar coordinates $(0, 0)$, $(2, \pi/6)$, $(-2, \pi/2)$, $(2, 5\pi/6)$.

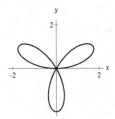

57. $r = 1 + 2\cos\theta$ goes through the following points in polar coordinates $(3, 0)$, $(2, \pi/3)$, $(1, \pi/2)$, $(0, 2\pi/3)$, $(1 - \sqrt{3}, 5\pi/6)$, $(-1, \pi)$.

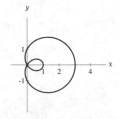

59. $r = 3.5$ is a circle centered at the origin with radius 3.5.

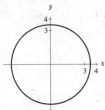

61. $\theta = 30°$ is a line through the origin that makes a $30°$ angle with the positive x-axis.

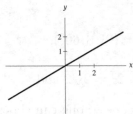

63. Multiply given equation by r.
$$\begin{aligned} r^2 &= 4r\cos\theta \\ x^2 + y^2 &= 4x \\ x^2 - 4x + y^2 &= 0 \end{aligned}$$

65. Multiply equation by $\sin\theta$.
$$\begin{aligned} r\sin\theta &= 3 \\ y &= 3 \end{aligned}$$

67. Multiply equation by $\cos\theta$.
$$\begin{aligned} r\cos\theta &= 3 \\ x &= 3 \end{aligned}$$

69. Since $r = 5$, we get $r = \sqrt{x^2 + y^2} = 5$. Squaring, we find $x^2 + y^2 = 25$.

71. Note, $\tan\theta = \dfrac{\sin\theta}{\cos\theta} = \dfrac{y/r}{x/r} = \dfrac{y}{x}$.
Since $\tan\pi/4 = 1$, $\dfrac{y}{x} = 1$ or $y = x$.

73. Multiply equation by $1 - \sin\theta$.
$$\begin{aligned} r(1 - \sin\theta) &= 2 \\ r - r\sin\theta &= 2 \\ r - y &= 2 \\ \pm\sqrt{x^2 + y^2} &= y + 2 \\ x^2 + y^2 &= y^2 + 4y + 4 \\ x^2 - 4y &= 4 \end{aligned}$$

75. $r\cos\theta = 4$

77. Note $\tan\theta = y/x$.

$$
\begin{aligned}
y &= -x \\
\frac{y}{x} &= -1 \\
\tan\theta &= -1 \\
\theta &= -\pi/4
\end{aligned}
$$

79. Note $x = r\cos\theta$ and $y = r\sin\theta$.

$$
\begin{aligned}
(r\cos\theta)^2 &= 4r\sin\theta \\
r^2\cos^2\theta &= 4r\sin\theta \\
r &= \frac{4\sin\theta}{\cos^2\theta} \\
r &= 4\tan\theta\sec\theta
\end{aligned}
$$

81. $r = 2$

83. Note $x = r\cos\theta$ and $y = r\sin\theta$.

$$
\begin{aligned}
y &= 2x - 1 \\
r\sin\theta &= 2r\cos\theta - 1 \\
r(\sin\theta - 2\cos\theta) &= -1 \\
r(2\cos\theta - \sin\theta) &= 1 \\
r &= \frac{1}{2\cos\theta - \sin\theta}.
\end{aligned}
$$

85. Note that $y = r\sin\theta$ and $x^2 + y^2 = r^2$.

$$
\begin{aligned}
x^2 + (y^2 - 2y + 1) &= 1 \\
x^2 + y^2 - 2y &= 0 \\
r^2 - 2r\sin\theta &= 0 \\
r^2 &= 2r\sin\theta \\
r &= 2\sin\theta
\end{aligned}
$$

87. There are six points of intersection and in polar coordinates these are approximately $(1, 0.17)$, $(1, 2.27)$, $(1, 4.36)$, $(1, 0.87)$, $(1, 2.97)$, $(1, 5.06)$

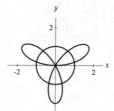

89. There are seven points of intersection and in polar coordinates these are approximately $(0, 0)$, $(0.9, 1.4)$, $(1.2, 1.8)$, $(1.9, 2.8)$, $(1.9, 3.5)$, $(1.2, 4.5)$, $(0.9, 4.9)$

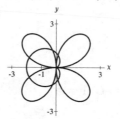

91. If the coordinate system is centered at A, then polar coordinates of B is $(1, 90°)$.

If the coordinate system is centered at B, then polar coordinates of C is $(\sqrt{2}, -45°)$.

If the coordinate system is centered at C, then polar coordinates of A is $(1, 180°)$.

93. Consider the trapezoid below.

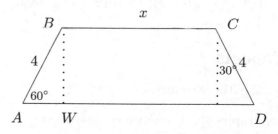

Note, $AW = 4\cos 60° = 2$. Since $AD = 10$, we find $x = 6$.

Thus, if the coordinate system is centered at A, then polar coordinates of B is $(4, 60°)$.

If the coordinate system is centered at B, then polar coordinates of C is $(6, 0°)$.

If the coordinate system is centered at C, then polar coordinates of D is $(4, -60°)$.

If the coordinate system is centered at D, then polar coordinates of A is $(10, 180°)$.

97. $\left(\sqrt{2}\left(\cos\dfrac{\pi}{4} + i\cos\dfrac{\pi}{4}\right)\right)^{12} =$

$64(\cos 3\pi + i\sin 3\pi) = -64$

99. **a)** $16 + 24i + 9i^2 = 7 + 24i$

b) $\dfrac{2+i}{4-i} \cdot \dfrac{4+i}{4+i} = \dfrac{7+6i}{17} = \dfrac{7}{17} + \dfrac{6}{17}i$

c) $-3 + 5i + 4 + 2i = 1 + 7i$

101. a) $[-3+1, 3+1]$ or $[-2, 4]$

b) Range $[1, \infty)$ for the parabola opens up from vertex $(0, 1)$

c) $(-\infty, \infty)$

d) $(-\infty, -2] \cup [2, \infty)$

103. Note, we have the sides

$$CT = CA = 5$$

and

$$CS = CB = 6.$$

Then $\triangle ABC$ is congruent to $\triangle TSC$. We use Heron's formula, and let

$$s = \frac{5+6+7}{2} = 9.$$

The area of $\triangle TSC$ is

$$\text{Area} = \sqrt{9(9-5)(9-6)(9-7)} = 6\sqrt{6}.$$

For Thought

1. False, t is the parameter.

2. True, graphs of parametric equations are sketched in a rectangular coordinate system in this book.

3. True, since $2x = t$ and $y = 2t + 1 = 2(2x) + 1 = 4x + 1$.

4. False, it is a circle of radius 1.

5. True, since if one substitutes $x = \tan t$ in $y = \tan^2 t$ one gets $y = x^2$. Note, the range of $y = \tan t$ is (∞, ∞).

6. True, since if $t = \dfrac{1}{3}$ then $x = 3\left(\dfrac{1}{3}\right) + 1 = 2$ and $y = 6\left(\dfrac{1}{3}\right) - 1 = 1$.

7. False, for if $w^2 - 3 = 1$ then $w = \pm 2$ and this does not satisfy $-2 < w < 2$.

8. False, since some of the points lie in the 3rd quadrant.

9. True, since $x = -\sin(t) < 0$ and $y = \cos(t) > 0$ for $0 < t < \pi/2$.

10. True, since $x = r\cos\theta$ and $y = r\sin\theta$ where $r = \cos\theta$ is given.

6.5 Exercises

1. parametric

3. If $t = 0$, then $x = 4(0) + 1 = 1$ and $y = 0 - 2 = -2$. If $t = 1$, then $x = 4(1) + 1 = 5$ and $y = 1 - 2 = -1$.

If $x = 7$, then $7 = 4t + 1$. Solving for t, we get $t = 1.5$. Substitute $t = 1.5$ into $y = t - 2$. Then $y = 1.5 - 2 = -0.5$.

If $y = 1$, then $1 = t - 2$. Solving for t, we get $t = 3$. Consequently $x = 4(3) + 1 = 13$.

We tabulate the results as follows.

t	x	y
0	1	-2
1	5	-1
1.5	7	-0.5
3	13	1

5. If $t = 1$, then $x = 1^2 = 1$ and $y = 3(1) - 1 = 2$. If $t = 2.5$, then $x = (2.5)^2 = 6.25$ and $y = 3(2.5) - 1 = 6.5$.

If $x = 5$, then $5 = t^2$ and $t = \sqrt{5}$. Consequently, $y = 3\sqrt{5} - 1$.

If $y = 11$, then $11 = 3t - 1$. Solving for t, we get $t = 4$. Consequently $x = 4^2 = 16$.

If $x = 25$, then $25 = t^2$ and $t = 5$. Consequently, $y = 3(5) - 1 = 14$.

We tabulate the results as follows.

t	x	y
1	1	2
2.5	6.25	6.5
$\sqrt{5}$	5	$3\sqrt{5} - 1$
4	16	11
5	25	14

7.

Some points are given by

t	x	y
0	-2	3
4	10	7

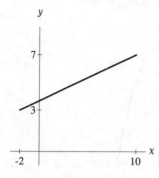

9.

Some points are given by

t	x	y
0	-1	0
2	1	4

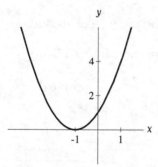

11.

A few points are approximated by

t	x	y
0.2	0.4	0.9
0.8	0.9	0.4

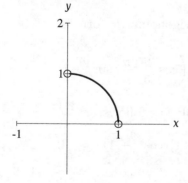

13. A circle of radius 1, centered at the origin

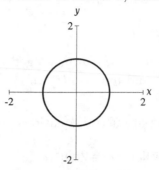

15. Since $t = \dfrac{x+5}{4}$, we obtain

$$y = 3 - 4\left(\frac{x+5}{4}\right) = -x - 2$$

or equivalently

$$x + y = -2.$$

The graph is a straight line.

17. Since $x^2 + y^2 = 16\sin^2(3t) + 16\cos^2(3t) = 16$, we get

$$x^2 + y^2 = 16.$$

The graph is a circle with radius 4 and with center at the origin.

19. Since $t = x - 4$, we find

$$y = \sqrt{(x-4) - 5} = \sqrt{x - 9}.$$

The graph of

$$y = \sqrt{x - 9}$$

is a square root curve.

21. $y = 2x + 3$ is the graph of a straight line

23. An equation (in terms of x and y) of the line through $(2,3)$ and $(5,9)$ is $y = 2x - 1$.

An equation (in terms of t and x) of the line through $(0,2)$ and $(2,5)$ is $x = \dfrac{3}{2}t + 2$.

The parametric equations are

$$x = \frac{3}{2}t + 2$$

and

$$y = 2\left(\frac{3}{2}t + 2\right) - 1 = 3t + 3$$

where $0 \le t \le 2$.

25. $x = 2\cos(t)$, $y = 2\sin(t)$, $\pi < t < \dfrac{3\pi}{2}$

27. $x = 3$, $y = t$, $-\infty < t < \infty$

29. Since $x = r\cos\theta$ and $y = r\sin\theta$, we get $x = 2\sin(t)\cos(t)$ and $y = 2\sin(t)\sin(t)$ where $0 \le t \le 2\pi$.

Equivalently, the parametric equations are

$$x = \sin(2t), \quad y = 2\sin^2(t)$$

where $0 \le t \le 2\pi$.

31. For $-\pi \le t \le \pi$, one obtains the given graph (for a larger range of values for t, more points are filled and the graph would look different)

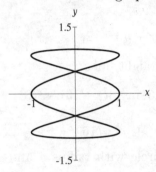

33. For $-15 \le t \le 15$, one finds

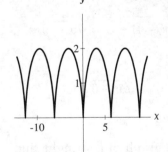

35. For $-10 \le t \le 10$, one obtains

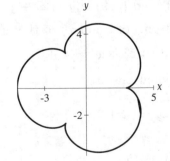

37. The graph of the parametric equations

$$x = 150\sqrt{3}\,t, \quad y = -16t^2 + 150t + 5$$

for $0 \le t \le 10$ is given below

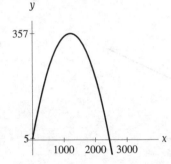

39. Solving $y = -16t^2 + 150t + 5 = 0$, one finds

$$t = \frac{-150 \pm \sqrt{150^2 - 4(-16)(5)}}{-32} \approx 9.41, -0.03$$

The arrow is in the air for 9.4 seconds.

41. Multiply given equation by r. Then

$$
\begin{aligned}
r^2 &= 8r\cos\theta \\
x^2 + y^2 &= 8x \\
x^2 - 8x + y^2 &= 0.
\end{aligned}
$$

43. Note, $|-3 - 3i\sqrt{3}| = \sqrt{(-3)^2 + (-3\sqrt{3})^2}$ $= \sqrt{9 + 27} = 6$. An argument in the fourth quadrant is

$$\arctan\left(\frac{-3\sqrt{3}}{3}\right) = -\frac{\pi}{3}$$

or $\dfrac{5\pi}{3}$. The trigonometric form is

$$6\left(\cos\frac{5\pi}{3} + i\sin\frac{5\pi}{3}\right).$$

45. Factor and solve as follows:

$$
\begin{aligned}
(2\cos x - 1)(\cos x + 1) &= 0 \\
\cos x &= \frac{1}{2}, -1 \\
x &= \frac{\pi}{3}, \pi, \frac{5\pi}{3}.
\end{aligned}
$$

47. In the figure below, we have $AB = 1$, $AC = x$, and $BC = \sqrt{1 - x^2}$. Let y be the length of the statue.

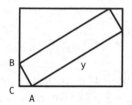

By the Pythagorean Theorem, we get

$$(4 - x)^2 + (3 - \sqrt{1 - x^2})^2 = y^2.$$

Using the area of the rectangle, we find

$$12 = x\sqrt{1 - x^2} + y + (4 - x)(3 - \sqrt{1 - x^2}).$$

The solutions of the system of two equations are

$$x \approx 0.5262 \text{ ft}, y \approx 4.0851 \text{ ft}.$$

The length of the statue is $y \approx 4.0851$ ft.

Chapter 6 Review Exercises

1. $-1 - i$

3. $(4 - 5i)^2 = 16 - 40i + (25i^2) = -9 - 40i$

5. $(1-3i)2 - (1-3i)6i = 2 - 6i - 6i - 18 = -16 - 12i$

7. $\dfrac{2 - 3i}{i} \cdot \dfrac{-i}{-i} = \dfrac{-3 - 2i}{1} = -3 - 2i$

9. $\dfrac{1 + i}{2 - 3i} \cdot \dfrac{2 + 3i}{2 + 3i} = \dfrac{-1 + 5i}{13} = -\dfrac{1}{13} + \dfrac{5}{13}i$

11. $\dfrac{6 + 2i\sqrt{2}}{2} = 3 + i\sqrt{2}$

13. $i^{32}i^2 + i^8 i = i^2 + i = -1 + i$

15. $|3 - 5i| = \sqrt{3^2 + (-5)^2} = \sqrt{34}$

17. $|\sqrt{5} + i\sqrt{3}| = \sqrt{(\sqrt{5})^2 + (\sqrt{3})^2} = \sqrt{8} = 2\sqrt{2}$

19. Note $\sqrt{(-4.2)^2 + (4.2)^2} \approx 5.94$. If the terminal side of α goes through $(-4.2, 4.2)$, then $\tan \alpha = -1$. If $\alpha = 135°$, then

$$-4.2 + 4.2i \approx 5.94 \left[\cos 135° + i \sin 135° \right].$$

21. Note $\sqrt{(-2.3)^2 + (-7.2)^2} \approx 7.6$. If the terminal side of α goes through $(-2.3, -7.2)$ and since $\tan^{-1}(7.2/2.3) \approx 72.3°$, then one can choose $\alpha = 180° + 72.3° = 252.3°$. Thus,

$$-2.3 - 7.2i \approx 7.6 \left[\cos 252.3° + i \sin 252.3° \right].$$

23. $\sqrt{3} \left(-\dfrac{\sqrt{3}}{2} + i\dfrac{1}{2} \right) = -\dfrac{3}{2} + \dfrac{\sqrt{3}}{2}i$

25. $6.5[0.8377 + (0.5461)i] \approx 5.4 + 3.5i$

27. Since $z_1 = 2.5\sqrt{2} \left[\cos 45° + i \sin 45° \right]$ and $z_2 = 3\sqrt{2} \left[\cos 225° + i \sin 225° \right]$, we have

$$z_1 z_2 = 15 \left[\cos 270° + i \sin 270° \right] = -15i$$

and

$$\dfrac{z_1}{z_2} = \dfrac{(5/2)\sqrt{2}}{3\sqrt{2}} \left[\cos(-180°) + i \sin(-180°) \right]$$

$$= -\dfrac{5}{6}.$$

29. Let α and β be angles whose terminal sides go through $(2, 1)$ and $(3, -2)$, respectively. Since $|2 + i| = \sqrt{5}$ and $|3 - 2i| = \sqrt{13}$, we get $\cos \alpha = 2/\sqrt{5}$, $\sin \alpha = 1/\sqrt{5}$, $\cos \beta = 3/\sqrt{13}$, and $\sin \beta = -2/\sqrt{13}$. From the sum and difference identities, we obtain

$$\cos(\alpha + \beta) = \dfrac{8}{\sqrt{65}}, \sin(\alpha + \beta) = -\dfrac{1}{\sqrt{65}},$$

$$\cos(\alpha - \beta) = \dfrac{4}{\sqrt{65}}, \sin(\alpha - \beta) = \dfrac{7}{\sqrt{65}}.$$

Note $z_1 = \sqrt{5}(\cos \alpha + i \sin \alpha)$ and $z_2 = \sqrt{13}(\cos \beta + i \sin \beta)$. Then

$$z_1 z_2 = \sqrt{65} \left(\cos(\alpha + \beta) + i \sin(\alpha + \beta) \right)$$

$$= \sqrt{65} \left(\dfrac{8}{\sqrt{65}} - i \cdot \dfrac{1}{\sqrt{65}} \right)$$

$$z_1 z_2 = 8 - i$$

and

$$\dfrac{z_1}{z_2} = \dfrac{\sqrt{5}}{\sqrt{13}} \left(\cos(\alpha - \beta) + i \sin(\alpha - \beta) \right)$$

$$= \dfrac{\sqrt{65}}{13} \left(\dfrac{4}{\sqrt{65}} + i \cdot \dfrac{7}{\sqrt{65}} \right)$$

$$\dfrac{z_1}{z_2} = \dfrac{4}{13} + \dfrac{7}{13}i \approx 0.31 + 0.54i.$$

31. $2^3 \left[\cos 135° + i \sin 135°\right] =$

$$8 \left[-\frac{\sqrt{2}}{2} + i\frac{\sqrt{2}}{2}\right] = -4\sqrt{2} + 4\sqrt{2}i$$

33. $(4 + 4i)^3 = (4\sqrt{2})^3 \left[\cos 45° + i \sin 45°\right]^3 =$
$128\sqrt{2} \left[\cos 135° + i \sin 135°\right] =$
$$128\sqrt{2} \left[-\frac{1}{\sqrt{2}} + i\frac{1}{\sqrt{2}}\right] = -128 + 128i$$

35. Square roots of i are given by

$$\cos\left(\frac{90° + k360°}{2}\right) + i \sin\left(\frac{90° + k360°}{2}\right) =$$
$$\cos(45° + k \cdot 180°) + i \sin(45° + k \cdot 180°)$$

where k is an integer.

If $k = 0, 1$ one gets

$$\cos 45° + i \sin 45° = \frac{\sqrt{2}}{2} + i\frac{\sqrt{2}}{2} \text{ and}$$
$$\cos 225° + i \sin 225° = -\frac{\sqrt{2}}{2} - i\frac{\sqrt{2}}{2}.$$

37. Since $|\sqrt{3} + i| = 2$, the cube roots are

$$\sqrt[3]{2} \left[\cos\left(\frac{30° + k360°}{3}\right) + i \sin\left(\frac{30° + k360°}{3}\right)\right]$$

where k is an integer. If $k = 0, 1, 2$, we find

$$\sqrt[3]{2} \left[\cos \alpha + i \sin \alpha\right]$$

where $\alpha = 10°, 130°, 250°$.

39. Since $|2 + i| = \sqrt{5}$ and $\tan^{-1}(1/2) \approx 26.6°$, the arguments of the cube roots are

$$\frac{26.6° + k360°}{3} \approx 8.9° + k120°$$

where k is an integer.

If $k = 0, 1, 2$, we obtain

$$\sqrt[6]{5} \left[\cos \alpha + i \sin \alpha\right]$$

where $\alpha = 8.9°, 128.9°, 248.9°$.

41. Since $\sqrt[4]{625} = 5$, the fourth roots of $625i$ are

$$5 \left[\cos\left(\frac{90° + k360°}{4}\right) + i \sin\left(\frac{90° + k360°}{4}\right)\right]$$

where k is an integer.

When $k = 0, 1, 2, 3$ the fourth roots are

$$5 \left[\cos \alpha + i \sin \alpha\right]$$

where $\alpha = 22.5°, 112.5°, 202.5°, 292.5°$.

43. $(5\cos 60°, 5\sin 60°) = \left(\dfrac{5}{2}, \dfrac{5\sqrt{3}}{2}\right)$

45. $(\sqrt{3}\cos 100°, \sqrt{3}\sin 100°) \approx (-0.3, 1.7)$

47. Note, we have

$$r = \sqrt{(-2)^2 + (-2\sqrt{3})^2} = \sqrt{16} = 4.$$

Since $\tan \theta = \sqrt{3}$ and the terminal side of θ goes through $(-2, -2\sqrt{3})$, we may let $\theta = 4\pi/3$. Then

$$(r, \theta) = \left(4, \frac{4\pi}{3}\right).$$

49. Note, the magnitude is

$$r = \sqrt{2^2 + (-3)^2} = \sqrt{13}.$$

Since $\theta = \tan^{-1}(-3/2) \approx -0.98$, we have

$$(r, \theta) \approx \left(\sqrt{13}, -0.98\right).$$

51. Circle centered at $(r, \theta) = (1, -\pi/2)$

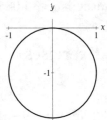

53. four-leaf rose

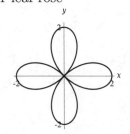

55. Limacon $r = 500 + \cos \theta$

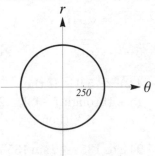

57. Horizontal line $y = 1$

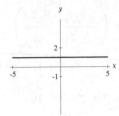

59. Since we have

$$r = \frac{1}{\sin\theta + \cos\theta},$$

we obtain

$$r\sin\theta + r\cos\theta = 1$$
$$y + x = 1.$$

61. $x^2 + y^2 = 25$

63. Since $y = 3$, we find $r\sin\theta = 3$ and $r = \dfrac{3}{\sin\theta}$.

65. $r = 7$

67. The boundary points are given by

t	x	y
0	0	3
1	3	2

Note, the boundary points do not lie on the graph.

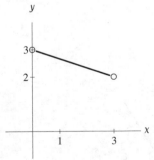

69. The graph is a quarter of a circle.

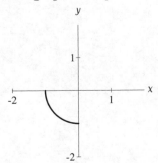

71. In the figure below,

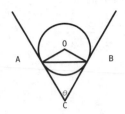

the angle at the center O is

$$\langle AOB = \pi - \theta.$$

Then the area of triangle $\triangle AOB$ is

$$A_t = \frac{\sin\theta}{2}$$

and the area of sector $\widetilde{AB}$ is

$$A_s = \frac{\pi - \theta}{2}.$$

a) The area inside the circle and below line AB is

$$A_s - A_t = \frac{\pi}{2} - \frac{\theta}{2} - \frac{\sin\theta}{2}$$
$$= \frac{\pi}{2} - \frac{\theta}{2} - \sin(\theta/2)\sin(\theta/2)$$

b) Applying the Pythagorean Theorem to $\triangle AOC$, we find

$$AC = 1/\tan(\theta/2).$$

Then the area of $\triangle AOC$ is

$$\frac{1}{2\tan(\theta/2)}$$

and the area of the quadrilateral $ABCO$ is

$$A_q = \frac{1}{\tan(\theta/2)}.$$

Hence, the area below the circle and inside the trench is

$$A_q - A_s = \frac{1}{\tan(\theta/2)} - \frac{\pi}{2} + \frac{\theta}{2}$$

c) Using a calculator, we find that as θ approaches π, the ratio

$$\frac{A_s - A_t}{A_q - A_s}$$

approaches 2.

Chapter 6 Test

1. $(4 - 3i)^2 = 16 - 24i + (3i)^2 = 7 - 24i$

2. $\dfrac{2 - i}{3 + i} \cdot \dfrac{3 - i}{3 - i} = \dfrac{5 - 5i}{10} = \dfrac{1}{2} - \dfrac{1}{2}i$

3. $i^4 i^2 - i^{32} i^3 = i^2 - i^3 =$

$\quad -1 - (-i) = -1 + i$

4. $2i\sqrt{2}(i\sqrt{2} + \sqrt{6}) = -4 + 2i\sqrt{12} = -4 + 4i\sqrt{3}$

5. Since $|3 + 3i| = 3\sqrt{2}$ and $\tan^{-1}(3/3) = 45°$, we have

$$3 + 3i = 3\sqrt{2}\left[\cos 45° + i\sin 45°\right].$$

6. Since $|-1 + i\sqrt{3}| = 2$ and $\cos^{-1}(-1/2) = 120°$, we have

$$|-1 + i\sqrt{3}| = 2\left[\cos 120° + i\sin 120°\right].$$

7. Note $|-4 - 2i| = \sqrt{20} = 2\sqrt{5}$.
Since $\tan^{-1}(2/4) = 26.6°$, the direction angle of $-4 - 2i$ is

$$180° + 26.6° = 206.6°.$$

Then we obtain

$$-4 - 2i = 2\sqrt{5}\left[\cos 206.6° + i\sin 206.6°\right].$$

8. $6\left[\cos 45° + i\sin 45°\right] =$

$\quad 6\left[\dfrac{\sqrt{2}}{2} + i\dfrac{\sqrt{2}}{2}\right] = 3\sqrt{2} + 3i\sqrt{2}$

9. $2^9\left[\cos 90° + i\sin 90°\right] = 512\left[0 + i\right] = 512i$

10. $\dfrac{3}{2}\left[\cos 45° + i\sin 45°\right] =$

$\quad \dfrac{3}{2}\left[\dfrac{\sqrt{2}}{2} + i\dfrac{\sqrt{2}}{2}\right] = \dfrac{3\sqrt{2}}{4} + \dfrac{3\sqrt{2}}{4}i$

11. $(5\cos 30°, 5\sin 30°) = \left(5 \cdot \dfrac{\sqrt{3}}{2}, 5 \cdot \dfrac{1}{2}\right) =$

$\quad \left(\dfrac{5\sqrt{3}}{2}, \dfrac{5}{2}\right)$

12. $(-3\cos(-\pi/4), -3\sin(-\pi/4)) =$

$\quad \left(-3 \cdot \dfrac{\sqrt{2}}{2}, 3 \cdot \dfrac{\sqrt{2}}{2}\right) = \left(-\dfrac{3\sqrt{2}}{2}, \dfrac{3\sqrt{2}}{2}\right)$

13. $(33\cos 217°, 33\sin 217°) \approx (-26.4, -19.9)$

14. Circle of radius $5/2$ with center at $(r, \theta) = (5/2, 0)$

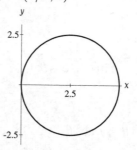

15. Four-leaf rose.

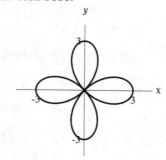

16. Some points on the graph are given by

t	x	y
-1	-2	-10
3	6	6

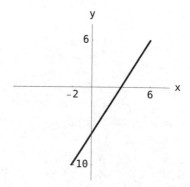

17. Some points on the graph are given by

t	x	y
0	0	3
π	0	-3

18. Fourth roots of -81 are given by

$$3\left[\cos\left(\frac{180° + k360°}{4}\right) + i\sin\left(\frac{180° + k360°}{4}\right)\right]$$
$$= 3\left[\cos\left(45° + k90°\right) + i\sin\left(45° + k90°\right)\right].$$

When $k = 0, 1, 2, 3$ one gets

$$3\left[\cos 45° + i\sin 45°\right] = \frac{3\sqrt{2}}{2} + \frac{3\sqrt{2}}{2}i,$$

$$3\left[\cos 135° + i\sin 135°\right] = -\frac{3\sqrt{2}}{2} + \frac{3\sqrt{2}}{2}i,$$

$$3\left[\cos 225° + i\sin 225°\right] = -\frac{3\sqrt{2}}{2} - \frac{3\sqrt{2}}{2}i,$$

$$3\left[\cos 315° + i\sin 315°\right] = \frac{3\sqrt{2}}{2} - \frac{3\sqrt{2}}{2}i.$$

19. Since $x^2 + y^2 + 5y = 0$, we obtain

$$\begin{aligned} r^2 + 5r\sin\theta &= 0 \\ r + 5\sin\theta &= 0 \\ r &= -5\sin\theta \end{aligned}$$

20. Since $r = 5(2\sin\theta\cos\theta)$, we find

$$\begin{aligned} r^3 &= 10(r\sin\theta)(r\cos\theta) \\ r^3 &= 10yx \\ (x^2 + y^2)^{3/2} &= 10yx. \end{aligned}$$

21. Since the slope of the line through $(-2, -3)$ and $(4, 5)$ is $\frac{5 + 3}{4 + 2}$ or $\frac{8}{6}$, parametric equations are $x = -2 + 6t$ and $y = -3 + 8t$ where $0 \le t \le 1$.

Tying It All Together

1. $\dfrac{\sqrt{2}}{2}$ **2.** $\dfrac{1}{2}$ **3.** $\dfrac{\pi}{4}$ **4.** $-\dfrac{\pi}{4}$

5. $\dfrac{\pi}{3}$ **6.** $\dfrac{2\pi}{3}$

7. $\cos\left(\dfrac{\pi}{4}\right) = \dfrac{\sqrt{2}}{2}$ **8.** $\sin\left(\dfrac{-\pi}{4}\right) = -\dfrac{\sqrt{2}}{2}$

9. Let $\cos\theta = \dfrac{3}{5} = \dfrac{x}{r}$. Since $\cos^{-1}(3/5)$ is an angle in the 1st quadrant, we get $x = 3$, $r = 5$ and $y > 0$. Since $5^2 = 3^2 + y^2$, we find $y = 4$. Then

$$\sin\theta = \frac{y}{r} = \frac{4}{5}.$$

10. Let $\sin\theta = -\dfrac{3}{5} = \dfrac{y}{r}$. Since $\sin^{-1}\left(-\dfrac{3}{5}\right)$ is an angle in the 4th quadrant, we get $y = -3$, $r = 5$ and $x > 0$. Since $5^2 = x^2 + (-3)^2$, we get $x = 4$. Then $\tan\theta = \dfrac{y}{x} = -\dfrac{3}{4}$.

11. $\left\{x \mid x = \dfrac{\pi}{2} + 2k\pi \text{ where } k \text{ is an integer}\right\}$

12. $\{x \mid x = 2k\pi \text{ where } k \text{ is an integer}\}$

13. Since $\sin x = \pm 1$, the solution set is

$$\left\{x \mid x = \frac{\pi}{2} + 2k\pi \text{ or } x = \frac{3\pi}{2} + 2k\pi\right\}.$$

14. Since $\cos x = \pm 1$, the solution set is

$$\{x \mid x = k\pi \text{ where } k \text{ is an integer}\}.$$

15. Since $\sin x = \dfrac{1}{2}$, the solution set is

$$\left\{x \mid x = \frac{\pi}{6} + 2k\pi \text{ or } x = \frac{5\pi}{6} + 2k\pi\right\}.$$

16. Since $\cos x = \dfrac{1}{2}$, the solution set is

$$\left\{x \mid x = \frac{\pi}{3} + 2k\pi \text{ or } x = \frac{5\pi}{3} + 2k\pi\right\}.$$

17. Since $\sin(2x) = 1$, we obtain $2x = \dfrac{\pi}{2} + 2k\pi$. The solution set is

$$\left\{ x \mid x = \frac{\pi}{4} + k\pi \right\}.$$

18. Since $\sin^2 x + \cos^2 x = 1$ is an identity, the solution set is the set of all real numbers.

19. Note, $x \cdot \dfrac{\sqrt{3}}{2} = 1$. Solving for x, we get

$x = \dfrac{2}{\sqrt{3}}$. The solution set is $\left\{ \dfrac{2\sqrt{3}}{3} \right\}$.

20. Note, $x \cdot 1 + x \cdot \dfrac{\sqrt{3}}{2} = 1$. Factoring, we get

$x\left(1 + \dfrac{\sqrt{3}}{2}\right) = 1$ or $x\left(2 + \sqrt{3}\right) = 2$. Solving

for x, we find $x = \dfrac{2}{2 + \sqrt{3}}$. By rationalizing

the denominator, we find that the solution set
is $\left\{ 4 - 2\sqrt{3} \right\}$.

21. Note $\sin 2x = 2 \sin x \cos x$.

$$
\begin{aligned}
4 \sin x \cos x - 2 \cos x + 2 \sin x - 1 &= 0 \\
2 \cos x (2 \sin x - 1) + (2 \sin x - 1) &= 0 \\
(2 \cos x + 1)(2 \sin x - 1) &= 0
\end{aligned}
$$

Then $\cos x = -\dfrac{1}{2}$ or $\sin x = \dfrac{1}{2}$.

The solution set is $\left\{ x | x = \dfrac{\pi}{6} + 2k\pi, \right.$

$\left. x = \dfrac{5\pi}{6} + 2k\pi, x = \dfrac{2\pi}{3} + 2k\pi, x = \dfrac{4\pi}{3} + 2k\pi \right\}$.

22. First, we factor by grouping.

$$
\begin{aligned}
4x \sin x + 2 \sin x - 2x - 1 &= 0 \\
2 \sin x (2x + 1) - (2x + 1) &= 0 \\
(2x + 1)(2 \sin x - 1) &= 0
\end{aligned}
$$

Then $x = -\dfrac{1}{2}$ or $\sin x = \dfrac{1}{2}$.
The solution set is

$$\left\{ x | x = -\frac{1}{2}, x = \frac{\pi}{6} + 2k\pi, x = \frac{5\pi}{6} + 2k\pi \right\}.$$

23. $y = \sin x$

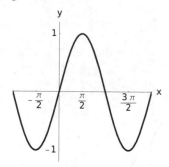

24. $y = \cos x$

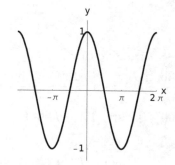

25. $r = \sin \theta$

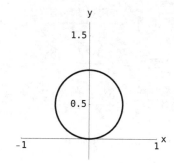

26. $r = \cos \theta$

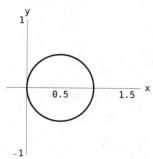

27. $r = \theta$

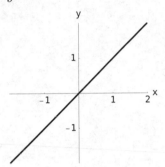

31. $y = \sqrt{\sin x}$

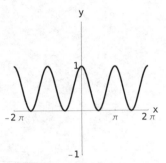

28. $y = x$

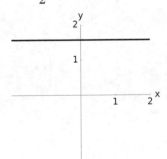

32. $y = \cos^2 x$

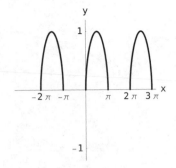

29. $y = \dfrac{\pi}{2}$

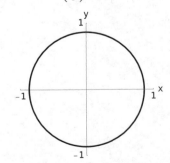

33. Heron's area formula

34. vector

35. scalar

36. magnitude

37. resultant

38. dot product

39. orthogonal

40. complex number

41. complex conjugates

42. De Moivre's

30. $r = \sin\left(\dfrac{\pi}{3}\right)$

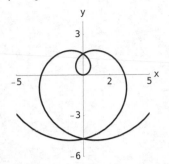